# Western Mainstream Media and the Ukraine Crisis



This book explores contemporary propaganda and mainstream Western news media, with reference to the Ukraine crisis.

It examines Western media narratives of the immediate causes of the crisis, the respective roles of those who participated in or otherwise supported the demonstrations of 2013–2014 – including US-backed NGOs and rightist militia – and the legitimacy, or otherwise, of the destabilization of the democratically elected Yanukovych government. It considers how the crisis was contextualized with reference to broader themes of competition for power over Eurasia and the Washington Consensus. It assesses accounts of the role of Russia and of ethnic Russian Ukrainians in Crimea, Odessa and the Donbass and traces how Western mainstream media went out of their way to demonize Vladimir Putin. The book deconstructs prevailing Western narratives as to the reasons for the shooting down of Malaysian Airways flight MH17 in July 2014, and counters Western media concentration on the issue of culpability for the attack with an alternative narrative of egregious failure to close down civilian air space over war zones. From analysis of these discourses, the book identifies principles of post-2001 Western conflict propaganda as these appeared to play out in Ukraine.

This book will be of much interest to students of propaganda, media and communication studies, Russian and Eastern European politics, security studies and IR.

**Oliver Boyd-Barrett** was formerly Director of the School of Media and Communications at Bowling Green State University, Ohio, USA, and now lectures at California State University. He is author of several books, including *Hollywood and the CIA: Cinema, Defense and Subversion* (Routledge 2011).

**Media, War and Security**

Series Editors: Andrew Hoskins
*University of Glasgow*
and
Oliver Boyd-Barrett
*Bowling Green State University*

This series interrogates and illuminates the mutually shaping relationship between war and media as transformative of contemporary society, politics and culture.

**Global Terrorism and New Media**
The post Al-Qaeda generation
*Philip Seib and Dana M. Janabek*

**Radicalisation and the Media**
Legitimising violence in the new media
*Akil N. Awan, Andrew Hoskins and Ben O'Loughlin*

**Hollywood and the CIA**
Cinema, defense and subversion
*Oliver Boyd-Barrett, David Herrera and Jim Baumann*

**Violence and War in Culture and the Media**
*Athina Karatzogianni*

**Military Media Management**
Negotiating the 'front' line in mediatized war
*Sarah Maltby*

**Icons of War and Terror**
Media images in an age of international risk
*Edited by John Tulloch and R. Warwick Blood*

**Memory, Conflict and New Media**
Web wars in post-socialist states
*Edited by Julie Fedor, Ellen Rutten and Vera Zvereva*

**Violent Extremism Online**
New perspectives on terrorism and the internet
*Edited by Anne Aly, Stuart Macdonald, Lee Jarvis and Thomas M. Chen*

**Western Mainstream Media and the Ukraine Crisis**
A study in conflict propaganda
*Oliver Boyd-Barrett*

# Western Mainstream Media and the Ukraine Crisis

A study in conflict propaganda

**Oliver Boyd-Barrett**

Routledge
Taylor & Francis Group
LONDON AND NEW YORK

First published 2017 by Routledge

2 Park Square, Milton Park, Abingdon, Oxfordshire OX14 4RN
711 Third Avenue, New York, NY 10017

*Routledge is an imprint of the Taylor & Francis Group, an informa business*

First issued in paperback 2018



*British Library Cataloguing-in-Publication Data*
A catalogue record for this book is available from the British Library

*Library of Congress Cataloging-in-Publication Data*
Names: Boyd-Barrett, Oliver, author.
Title: Western mainstream media and the Ukraine Crisis : a study in conflict propaganda / Oliver Boyd-Barrett.
Description: New York, NY : Routledge, [2017] | Includes bibliographical references and index.
Identifiers: LCCN 2016016766| ISBN 9781138677197 (hardback) | ISBN 9781315559681 (ebook)
Subjects: LCSH: Ukraine Conflict, 2014—Mass media and the conflict. | Mass media–Political aspects–Ukraine. | Information warfare–Ukraine. | Communication in politics–Ukraine. | Mass media–Political aspects–Russia (Federation) | Communication in politics–Russia (Federation) | Propaganda–History–21st century. | Mass media–Political aspects–History–21st century. | Communication in politics–History–21st century.
Classification: LCC DK508.852 .B68 2016 | DDC 947.7086–dc23
LC record available at https://lccn.loc.gov/2016016766

ISBN: 978-1-138-67719-7 (hbk)
ISBN: 978-1-138-57989-7 (pbk)

Typeset in Times New Roman
by Wearset Ltd, Boldon, Tyne and Wear

**To Jack, Irene, Audrey, Leah, Claudia, Francesca, Jonathan, Daniel, Andrea, John, Sofia and Dean. From love through love to love.**

# Contents

# 1 Introduction

## Competing narratives of the Ukraine Crisis 2014–2015

My purpose in this book is to explore the implications for an understanding of contemporary propaganda and the role played in it of mainstream Western news media (WMM) with particular reference to the crisis which emerged in Ukraine following the forced change of government in February 2014, leading shortly thereafter to an escalation of tension between the new government of Kiev and the Ukrainian (but substantially ethnic Russian) regions of Crimea (which was shortly to vote to be annexed to Russia) and Eastern Ukraine, and between the United States of America (USA), the European Union (EU) and Russia each of whom professed to have important security interests in the unfolding of the crisis.

The principal focus throughout the study is on coverage of the Ukraine crisis by WMM, especially the prestige news organizations of the USA and UK, and how this coverage was challenged directly or by implication, by alternative, mainly Western media as well as by Russian and other media. Russian media or for that matter the media of the Donbass or of Crimea during this period are equally open to investigation for the propaganda functions they undoubtedly supplied, and it is important that studies be undertaken to make such investigations. Russian media are not, however, my main or overriding interest. Over a number of different publications, I have sought to better understand the nature of the relationship between centers of power in the West and WMM, especially following the events of 9/11 and the ensuing "war on terrorism." In previous publications, I have examined the WMM coverage of imperial wars that have been initiated on false or misleading pretexts. For reasons that will become clear, I believe that the Ukraine crisis was initiated and sustained to provide just such a pretext, should it be needed, in the context of escalating tensions between the nuclear powers of the USA and its allies in the European Union, in opposition to Russia and China.

I consider this topic to be important not only in itself but also because of how it might inform the study of communications in the context of a multi-polar world shared, on the one hand, between the USA and its traditional allies and vassal states and, on the other, the emerging alliances between the nations of the BRICS (Brazil, Russia, India, China and South Africa) economies. Notwithstanding a still powerful influence of US communications within the overall ambit of international media communication (Boyd-Barrett 2015), there is little evidence of a return flow from leading members of the BRICS alliance to the USA (e.g., Xie and Boyd-Barrett 2015). Analysis of media flows between the BRICS economies is in its infancy: extrapolating from what is known, such flows are likely to be relatively modest in volume, format range (e.g., exports of Brazilian telenovelas), and audience (e.g., exports of Indian and Chinese movies to Indian and Chinese diasporas).

The issue of flow may be less significant than the ways in which the content of international communications represent, reinforce or resist phenomena of imperialism or neo-imperialism (Boyd-Barrett 2015). The focus here on WMM representations of the Ukraine crisis speaks to US and Western attitudes and ambitions in relation to NATO expansion to Russia's borders (and what this may signify for the potential continuity of Western global hegemony) and how these in turn relate to perceptions of Russia as a potential aggressor, as a potential target (in the event of future fragmentation of the Russian Federation) or as a potential competitor in a struggle for influence over Eurasia, the continental landmass that separates Western Europe and China. In 2015, NATO had a membership of twenty-eight countries (with the USA footing 70 percent of the expenses) with several others, including Ukraine, vying for entry.

The one-sidedness of corporate mainstream Western media coverage in support of the official perceptions of Washington speaks not to the pluralism of a media era lauded by celebrants of digital and social media but to a much older narrative of complicity with the propaganda aims of imperial power. The Princeton scholar of Russia, Stephen F. Cohen, has observed that whereas during the Cold War "the media were open – the New York Times, the Washington Post – to debate," today "they no longer are. It's one hand clapping in our major newspapers and in our broadcast networks" (Smith 2015a). This reinforces trends that have been observed widely elsewhere, including in the study of Western media coverage of the (mainly) US invasion and occupation of Afghanistan in 2001, the (mainly) US invasion and occupation of Iraq in 2003, the disputed 2009 elections in Iran, unsupported and misleading Western claims that Iran's nuclear energy program at any time constituted a dangerous nuclear weapons threat, and the Western-supported destabilization of Libya from 2011, and of Syria from 2011.

There is little evidence here that the rise of the BRICS, and the supposedly countervailing power that a BRICS alliance (or alliances) might have in arresting the foreign policy ambitions of the USA and its allies have yet impacted WMM perspectives on the world or that a greater range of perspectives, putatively available as a result of the rise of BRICS media power, has helped shape Western media coverage in a more nuanced, multi-perspectival way.

In this book, I identify ten principal Manichean sub-plots to the Ukraine crisis. For each sub-plot, I examine the conflict between a narrative largely espoused by the Kiev regime and its US and EU partners and almost uniformly by WMM throughout the USA and Western Europe and the countries of their allies, on the one hand, and a counter-narrative constructed by many critical, alternative media in the West, on the other. The counter-narratives might sometimes favor perspectives emanating from Russian and ethnic Russian media and other sources but were generally not dependent on these. In making this determination I have been guided by my personal daily monitoring over 2014 and 2015 of: (1) selective Western, Russian and Ukrainian mainstream media (*Independent, Kyiv Post, Los Angeles Times, Moscow Times, New York Times, UNIAN Information Agency*); (2) selective Western alternative media (*Antiwar.com, Common Dreams, Consortium News, Information Clearing House, Huffington Post, Truthout.org*); and regular monitoring of (3) *Agence France-Presse, Bloomberg, BBC, Economist, Foreign Policy, Guardian, Reuters, RT, Sputnik*; and (4) *Fair.org., The Intercept, Global Research, MediaLens.org, Mint News, New Cold War* and *Vice News*. In the case of some alternative media in category (4), such as *Mint News* and *New Cold War*, I switched from regular to daily monitoring during the course of the study. I copied most substantial stories for later consideration.

From stories found on and copied from these websites over the period 2014–2015 that covered the Ukraine crisis or were directly relevant to it, I have constructed Weberian "ideal type" characterizations of the principal divergent narratives (Weber [1904] 1949). I focus particularly on Western media and what we can learn from these about the nature of Western conflict propaganda and the beliefs and presumptions that these invite of audiences amidst the exposure of hypocrisies and contradictions that propaganda so often entails. My methodology stems from grounded theory (Glaser and Strauss 1967). I work inductively from a range of texts that represent the clash of discourses between mainstream Western media accounts, on the one hand, and competing accounts that come principally from Western alternative media and companion sources that pay sustained attention, among other things, to the perspectives of ethnic Russians in Ukraine (Laclau and Mouffe 2014) on the other. Such a clash inevitably tends towards the destabilization of the hegemonic Western discourse, not in the sense that it entitles an analyst to declare what is "true" or "false," but in the sense of being able to detect the play of ideology amidst apparent contradiction, paradox and hypocrisy.

The ten main sub-plots or media narratives (some of them are several threads) have to do with the following episodes or phases of the conflict. I devote separate chapters to some of these, but I also bundle a few together, so that issues of International Monetary Fund (IMF) lending and sanctions against Russia, for example, are addressed within a broader conversation about context in Chapters 3 and 4. Here are the ten sub-plots and the Manichean narratives associated with each:

*1 Cause of the 2014 Crisis*

The West and Western media ascribed this largely to the yearning of the Ukrainian people to be identified with Europe rather than with a corrupt regime that had close ties to Russia. Russia and Russian media, East Ukrainian and alternative media in the West largely ascribed the crisis to a US-instigated coup that took place in close partnership with neo-fascist parties without regard for established, constitutional democratic procedures.

*2 Choice of IMF as source of financial aid*

The West and Western media represented the IMF as a benevolent savior that would restore the Ukrainian economy to robust and modern normalcy. Russia and alternative Western media regarded the IMF as a Pied Piper whose mission was to lure Ukraine into unbearable austerity while Western interests looted the economy and (further) reduced Ukraine to the status of a basket-case economy.

*3 New regime in Kiev*

The West and Western media represented the new regime as a legitimate response to authoritarian rule by a corrupt administration. Russian and alternative media in the West regarded the new regime as illegitimate, racist and anti-democratic, particularly in its "monist" disregard of the pluralist aspirations of the peoples of Ukraine's southern and eastern regions.

*4 The Maidan (described by its supporters as the EuroMaidan) movement*

The West and Western media represented the Maidan movement in much the same way as they had celebrated the various manifestations of the "Arab Spring" in 2011–2013, as one characterized by its youth, idealism and a thirst for modern, transparent democracy. Russian media and alternative Western media regarded Maidan as a movement that was instigated by faux "NGOs," such as the National Endowment for Democracy (NED) (funded by the US State Department), following a tried and tested "color revolution" script, in alliance with right-wing political and paramilitary movements in Ukraine, some of them neo-Nazi, and in alliance with Ukrainian oligarchs.

*5 Context*

The West and Western media offered relatively little context other than presenting Ukraine as a territory once ruled by the Soviet Union and, not least because of dependence on Russian energy, still in the icy grasp of the former Soviet Union and its allegedly expansionist ambitions. Russian and alternative Western media discerned in the obsessive neo-conservative US interest in Ukraine all the signs they needed that a key sector of the US elite wanted a war with Russia that would destabilize the Russian Federation and create a new pro-Western, capitalist Eurasian region that, in turn, would help contain China.

*6 Crimea, Odessa and Eastern Ukraine*

The West and Western media considered that Crimea had been forcibly seized by Russia and that Russia had assisted local ethnic Russian thugs in the Donbass to establish separatist fiefdoms in Luhansk and Donetsk. All these considerations amounted to legitimation for the imposition of US and EU economic sanctions against Russia. Russian and alternative Western media on the other hand stressed that the population of Crimea, following a long-established historical preference, had overwhelmingly voted to be annexed by Russia. In this perspective Russia had strategically maintained a very cool distance between itself and separatist movements in Eastern Ukraine, other than responding to urgent humanitarian needs and endeavoring to protect ethnic Russian populations from the worst of fascist cruelties, while seeking a political, federalist solution that would allow a strong measure of autonomy to the peoples of the Donbass region yet preserving it as part of Ukraine.

*7 Representing Ukraine and Russia to the world*

For the West and Western media, Ukraine – under its new leadership – could do no harm; it was an incontestable source of good. Russia on the other hand was a devilish source of evil, whose pinnacle was especially manifested in the words and actions of Russia's President Vladimir Putin. For Russia and Russian media, Ukraine was acknowledged as a sovereign state but one whose actions vis-à-vis the USA and EU had significant implications for Russian security, and in which Russia had legitimate interests and concerns. In this perspective the fact that the new Prime Minister, Arseniy Yatsenyuk, had not been elected (though Parliament confirmed him as prime minister in the September 2014 elections) and that the new president, billionaire oligarch Petro Poroshenko was the elected replacement of illegitimately deposed former President Viktor Yanukovych, hardly represented a triumph of democracy.

*8 Shooting down of MH17*

For the West and Western media, the shooting down of Malaysian Airlines civilian passenger jet MH17 was unmistakably and immediately recognizable as the work of ethnic Russian separatists, likely aided and abetted by Russia, possibly deliberate – even though an official Dutch-led investigation was not completed until sixteen months after the event and a criminal investigation report was not expected before mid-2017. Yet within weeks of its occurrence, WMM judged the downing to provide legitimate further justification for US and EU economic sanctions against Russia. For Russian and alternative Western media, the plane was likely shot down by Ukrainian military and was possibly a deliberate attempt by Ukraine and/or the West to smear the ethnic Russian separatists and Russia.

*9 Russia in Ukraine*

For the West, Western and Ukrainian media, Russia was periodically invading Ukraine or sending soldiers to fight alongside separatists and in this way justifying continuous US, NATO and EU concern and intervention, including the supply of military advisors, non-lethal and lethal arms. Ukrainian mainstream media routinely referred to fighters in Donbass as Russian "proxies." All this putative Russian activity provided further legitimation for the imposition of US and EU economic sanctions against Russia and helped NATO promote Russia as a potential threat to other Russian neighbors including the Baltic powers, Poland and Romania. For Russian and alternative Western media, Russia had never invaded the Crimea – since it legitimately stationed troops in the Crimea from before the crisis in defense of the Russian port of Sebastopol. Despite enormous provocation, Russia refrained from direct involvement in the Donbass. Even with the passage of time, with direct Russian admissions of an intelligence presence in Eastern Ukraine and of the presence of Russian volunteers, Western claims of a Russian "invasion" were mysteriously lacking in compelling official satellite or other hard evidence and much of whatever evidence there was originated from non-conventional and curious sources (such as Bellingcat.com) dependent on social media, or untrustworthy sources such as Ukrainian intelligence services.

*10 Sanctions against Russia*

For the West, Western and Ukrainian media, the application of sanctions against Russia was a justifiable, cautious, proportionate response to continuing Russian aggression against Ukraine. For Russian and alternative media, the application of sanctions represented an act of Western and Ukrainian aggression, based on false pretext (the false claim that Russia had "seized" Crimea, and "invaded" Ukraine), that was not only very damaging to the Russian economy (in addition to other measures, possibly manipulated, such as OPEC's maintenance of high levels of oil production amidst continuing deflation of oil prices) but was also counter-productive given that sanctions negatively impacted the economies of Western Europe, especially Germany.

In this book I shall be examining each of these narratives either individually or in combination, with a view to identifying, through close reading of both mainstream and alternative press accounts, what appear to be the principal propaganda intentions of the Western powers as represented by Washington, the EU and NATO and how these intentions are conveyed through WMM. When discussing movements within Ukraine that are opposed to the authorities in Kiev, I have generally referred to these as "separatist," although many opponents would be better described as federalist or pluralist, seeking not independence from Ukraine but a more equitable and autonomous status within it. While many may be ethnic Russians, and regard Russian as their first or predominant language, that does not necessarily mean that they are pro-Russia or anti-Ukraine.

## Introductory remarks on conflict propaganda

My intention in this study is not to reveal that any given set of texts is "biased." To undertake that kind of investigation would be to suggest that there are some sets of text that are not "biased." My position is that all communication, and by implication therefore all "texts" – the imprints of human communicative actions on the environment, paper, celluloid, airwaves, digital impressions, etc. that elicit perusal, with a view to the detection of meanings, by other humans – are "biased." This is because all human communication, in order to have communicative agency, must select, focus and highlight certain potentials for meaning at the expense of others. This is an inescapable feature of the nature of human sense-making, one that is engrained in human biology.

There is more scope for helpful debate as to whether some texts have propagandistic or persuasive quality or potential while others do not. If we start with the concept of persuasion, we could say that it requires intention, with a view either to altering given cognitions, attitudes or behaviors of others or to reinforcing or solidifying their existing cognitions, attitudes or behaviors. Or perhaps some combination of the two. But this postulate is more complex than at first seems to be the case. Communications are often if not usually motivated by several intentions, first of all, not all of them easy to decipher, often involving some mixture of relational intent (bonding or interacting with interlocutors) and the substantive or cognitive content of the communication. These are the products, therefore, of the identities of the parties doing the communicating, the nature of the relationships between them and the ostensible purposes of the communication within specific contexts and by means of specific linguistic and other communicative modalities. Intentions are not always readily accessible to the consciousness of those who own them but may be the result of unreflective habit or presumption. Successful persuasion may often entail forms of communication that begin intentionally and purposefully but that are later adopted by others whose level of reflexive awareness, if any, is lower. Yet these may continue to be diffused and have impact throughout a network, population or culture at a taken-for-granted level. In such processes of persuasion, therefore, a multitude of agents contributes at very different levels of awareness, rationality and intentionality.

If in theory the notion that there is generally a distinction between what is intended and what actually results may seem logical enough, it may be very difficult in practice for those involved or for a third party to determine the relationship between what happens and what was intended, or if what happens was intended at all. "What happens" is no straightforward entity but can be a complex array of cognitions, attitudes and behaviors that may or may not be explicit or visible, that may occur over time or at any particular time, and may be inextricably integrated into other communications involving the same or different parties that have taken place before, during or after the persuasive communication at issue. Since a communication can be persuasive in reinforcing existing cognitions, attitudes or behaviors, there may be no visible nor even measurable result of the persuasion at all.

The term propaganda suggests a more readily identifiable or specific form of communication than mere persuasion (for a good introduction and discussion of terms, see Jowett and O'Donnell 2015), although the terms used to define it sometimes appear arbitrary or insufficiently contained. Like persuasion, it is defined in many different ways. Jowett and O'Donnell suggest that unlike persuasion, propaganda has a more malign character, that in the case of propaganda the interests of the target of persuasion are not paramount, and may carry no weight whatsoever. Instead, it is the interests of the propagandist that constitute the driving forces of the communication, though it can help if the targets are persuaded that their interests coincide with those of the propagandist. What is "malign" or "benign" in the domain of interest can sometimes be obvious, yet oftentimes obscure, depending on how the term interest is defined, how interests may change over time or according to context, whether individual, group or some other collective form of interest is awarded precedence, and the extent to which interests are perceived or articulated by their owners. Nevertheless, in many definitions of propaganda there lies the imputation not just of malign or sinister interest at work, but, by contrast to mere persuasion, a greater likelihood of deliberate misinformation, even falsehood, or deception that may be based as much on the exclusion of certain categories of important and relevant information as on their inclusion. Other criteria may have to do with the degree of institutional involvement, the extent of deliberation, investment of resources, and evidence of systematic endeavor that involves many people and specialisms – possibly across many different individual communications over different platforms and maybe over extended periods of time.

Certainly, it is possible and probably desirable to delimit a definition of propaganda so as to mark it off from other forms of communication. This consideration helps explain several recent substitutes for the term in the military domain, such as strategic communication (SC) (Paul 2011), perception management (PM) and information operations (IO) (Armistead 2004). Each of these has the advantage of signaling not just a one-off message or even a full-blooded campaign (for example, to sell a war) so much as the deliberate construction of a totalistic information environment that will organically arrange, shape or contour mainstream definitions of reality, so that in turn these best engender the cognitive and affective conditions that the SC, PM or IO managers believe will most likely ensure their domination and control of the enemy. These concepts reach far beyond everyday understandings of communication-as-text to embrace "real" actions or actions that are constructed to seem "real" and that will reinforce the main goal. They might include military actions undertaken simply for symbolic value to demonstrate power and to instill fear or doubt in the minds of actual or potential future antagonists; false flag operations, sometimes to provide "atrocity stories" or images of atrocity that might not otherwise be easily available, often as pretexts for war; bribery of principals to create the appearance of authentic support; "Astroturf" or otherwise paid-for demonstrations of support for preferred candidates or policies; covert financing of preferred political parties; secret control over influential media or their employees; the supply and direction over the use of "props," as in the timely provision of US flags in the hands of supposedly overjoyed "liberated" people. In these kinds of operations, the use of public relations companies and actors is routine.

Not infrequently, even the manifest character of a propaganda campaign is a theatrical construction intended to distract public attention from the unspeakable. The "Cold War" was constructed as a struggle-to-the-death between incompatible ideologies – capitalism versus communism – and most of the people who engaged in that campaign, and those who were subject to it for over half a century, interpreted the Cold War solely in those terms. Its fundamental binary was a tragic intellectual simplification. It deliberately overlooked the different forms of capitalism, as between the competitive capitalism of Adam Smith and the monopolistic or oligopolistic capitalism of corporate USA, between regulated and unregulated capitalism, and the wide range of different models for an integration of capitalism with socialism, as in the forms of welfare capitalism in Western Europe. Even so, this titanic ideological battle was itself a curtain masking other interests. Some people have called it a war against the Third World (Dorrell 2005; Noble 2012; Perkins 2005), a pretext for the corporate grab of Third World resources through the calculated manufacture of indebtedness, the promise of dazzling modernization as a preferred alternative to what mainstream media depicted as the dreariness of socialism.

The so-called "war on drugs" may be thought of in similar terms as a propaganda campaign whose purposes included the camouflaging of a corporate grab for resource-rich land in areas of supposed drug cultivation; providing finance to US right-wing government puppets, their militaries and paramilitaries in Latin America; and diverting public monies in the construction of ready markets for Monsanto Roundup and other corporate products. The so-called "war-on-terror" equally may have camouflaged the waging of wars of false pretext against a string of countries that were not playing according to the preferred neo-conservative script of the Washington Consensus. These included Afghanistan, Iraq, Iran, Libya, Russia, Syria, Venezuela and Ukraine. Campaigns against them further diverted public monies to mainly private corporate interests of the military-industrial complex and fossil fuel industries. The supposed "restructuring" of the destabilized regimes that often resulted provided more opportunity for privatization, reconstruction and corporate looting. Propaganda campaigns with similar covert goals have included pro-democracy campaigns (whose purpose often is to shoe into power a local alliance of interests that is favorably disposed to or easily manipulated by imperial interest), humanitarian interventions, and alleged machination of fears of "weapons of mass destruction (WMD)" (most egregiously in the case of Iraq, 2003, which had already destroyed its WMD, and Iran, which never had such weapons nor is known even to have wanted them) (Boyd-Barrett 2015).

So much of perceived reality is the organic and dialectical product at some level or another of what were once propaganda or persuasion campaigns, formal or informal. Whatever their origin in the domains of governance, production, and religious, ethical, social class, gender, ethnic and other cultural relations, they are repetitively sustained through the constant invocation, for good or bad, through word and deed, of the postulates upon which the campaigns were initially based, even when these are generally perceived as normal, beyond question, taken-for-granted and especially when the contexts that gave rise to them, and the principles which they espoused, have been transformed beyond recognition. This is the case for example, with institutional, cultural and linguistic practices that we call "democratic" (even when they are not really, as in plutocratic "democracies," where all the props and paraphernalia of democratic processes cannot hide the vastly disproportionate influence wielded by people and institutions of power); "educational" (even when purchased at a level that enslaves the "educated" individual to lifelong debt, and is designed to foster a worker-citizen who will not challenge authority, while its institutions are increasingly contaminated by dependence on corporate sources for research and other funding); "healthy" (even when claims to "health" rest on controversial if not harmful practices such as GMO foods, carcinogenic herbicides, corn syrup); "legal" (where laws and their implementation are determined by those whose interests they serve); "defensive" (even when the "need" for "defense" is the result of provocation of the other); "productive" (even when the "productive" activity contributes to pollution, excessive $CO_2$ emissions and other contributions to planetary and species degradation); "humane" or "good" (as defined to suit the convenience of those who have the power to define the terms or identify their practice). There is a slipperiness, in short, about all the key terms attaching to ideas of persuasion and propaganda once it becomes apparent that the processes to which they contribute help constitute the very essence of social being. Such slipperiness resists facile bracketing-off for analytical purposes.

Since the invention of writing, there has been a close correspondence between propaganda, persuasion and what have come to be known as the mass media or, simply, the media. Media have greatly extended the reach of those who exercise control over and access to them – generally the same hands that exercised political, military, economic and religious power, usually party to shifting alliances between these four sources. Media power has been shared to only a modest degree with alternative or rival interests, reflecting the logic of business models (including advertising and subscription income) and the flexibility of technology innovations that have enhanced access to media for the purposes of both consumption and production. Universally, those with the power to ensure that their voices get to be the most heard and most disseminated in society continue to be those with the most means – through ownership, capital and legitimate authority – to run the media and determine their contents. Along with that privilege comes the entitlement to determine (on the basis of epistemologies and knowledge practices convenient to power) which "truths" should be societally endorsed and which modes of expression should be most celebrated (on the basis of technical capabilities and communication skills that routinely capture the attention of the greatest number of the people "who count" in the eyes of power).

The contents of media can no longer be separated off, if they ever could be, from their relations to the institutions and technologies which media use for the production, distribution and reception of content, or from more fundamental infrastructures of print, celluloid, cable, airwaves, satellite, computing and digital wireless – all of these representing significant industrial centers of capital accumulation. Defined most broadly, media serve power by: (1) providing the communications infrastructure that facilitate overall processes of capital accumulation, both theirs and those of their clients, in turn enabling international trade and contemporary globalization; (2) amplifying, giving maximum amplitude to the voices of the powerful, even as they give voice, though much less consistently, to the powerless and marginalized; and (3) stimulating the popular purchase of mass-produced media hardware and software products that feed the profitability of media enterprises. These in turn are closely intertwined: (4) through horizontal and vertical integration or conglomeration with major financial and industrial interests worldwide. Further, media serve power by: (5) acting as conduits for advertising that promotes the goods and services, including the goods and services of media industries themselves, that accumulate capital for the powerful. Through their hybridized informational and entertainment contents: (6) they disseminate ideas, values, ideologies that mainly promote consumption of capitalist goods and services, justify or normalize capitalism and the forms of "democracy" that are most compatible with it, and secure popular compliance with authority. And (7) media are fundamentally important to military and intelligence and to the growth of a surveillance society in the following ways: their information networks and programs provide open intelligence to the military; their contents generally serve the military interests of their host states; media technologies facilitate the ability of military and intelligence to "listen in" on a wide variety of publics; when integrated with weaponry such as robots and drones, media technologies become component parts of the war machine itself; the vast public and private resources invested in military hardware have also often brought about transformational advances in communications technologies available for commercial application (including satellite and the Internet).

Most media studies literature dealing with the relationships between media content and conflict has drawn from the empirical disciplines of psychology, social psychology and sociology as applied to persuasion and propaganda (Jowett and O'Donnell 2015; Glynn and Herbst 2015). Much of this has dwelt on media effects and the responses of media consumers to media contents, considering how and why consumers "read," "take meaning" from and interpret texts (Dickinson *et al.* 1998). From this tradition has come an appreciation of the significance of such factors as the interpersonal networks of media consumers and the role of opinion leaders; the perceived credibility of texts; the processes of selective exposure, attention and retention; and the significance of socio-economic, gender, ethnic, educational and other demographic details of audience members. Theories such as cognitive dissonance, the elaboration likelihood or social judgment model have helped identify the circumstances in which people are more or less likely to attend closely to given messages.

By paying attention to media as institutions, the study of media greatly extended our understanding of how media texts come to be. Gatekeeping studies of the press initiated an exploration of how information is funneled from the external world into the newsroom, newspapers and newscasts, and these in turn helped generate a broader interest in the range of different factors and variables that impinge on news selection – including policy, ownership and targeted audience, and the different weights that should be attributed to these (Shoemaker and Vos 2009; Franklin and Hamer 2005). Such considerations linked up with agenda-setting studies. These called attention to the impact of news agendas on the issues that the public considers to be most important at any given time. Agenda-setting research encouraged the exploration of the range of internal and external agencies that impact news agendas (including public and press relations agencies, and wire services) (McCombs 2014). The political economy "propaganda model" of Herman and Chomsky (2002) stressed five major variables of interest in the determination of news agendas: media ownership and corporate interest, heavy dependence on advertising for revenue, journalistic routines that privilege "authoritative" sources, journalists' fear of "flak" from powerful media sources; and uncritical ideological uniformity with the principle that – in a Cold War context – the Western way of life is superior to the communist or, since the end of the Cold War, that the "Washington Consensus" (a neo-conservative vision of global economic integration under US leadership) is "common sense."

The study of how media "frame" issues by consistently highlighting, emphasizing or prioritizing certain categories of information or angles of news, while backgrounding or disappearing others has also shown how readers or viewers are "primed" as to what to look for in candidates for political office (Entman 2003; Molden 2014). Concepts of "indexing" pick up on an older tradition of studies into the relationship between journalists and sources, noting in particular how journalists first seek out and privilege those whose narratives are closest to the major centers of power, later extending to outer circles of power particularly when there are significant areas of dissent within the power elite (Bennett *et al.* 2007).

Case studies of propaganda in conflict, sometimes written by participants, have proved important in exposing propaganda methodologies, both routine and exceptional. These include the admissions of George Creel, who had been the director of the US Committee on Public Opinion, as to the uses of propaganda by the USA in World War I (Creel 2012), many studies of Nazi propaganda, and Kermit Roosevelt's narrative of how the USA and Britain unseated the democratic Persian government in 1953 (Roosevelt 1979). From such studies have come attempts to standardize the essence of propaganda strategies as in those summarized by The Institute of Propaganda Analysis (1937–1942) (Jowett and O'Donnell 2015).

The overall outcome of these different literatures is to assemble, primarily by application of the scientific model, the basic principles, rules or patterns that can be identified across different manifestations of persuasion and propaganda. Powerful contemporary analyses include the documentary works of Adam Curtis (2002, 2004) and of Scott Noble (2012). Such attempts go further than many of their predecessors in identifying broader, underlying "mega-propagandas." These connect individuated, isolated examples of propaganda with the fundamental interests of competing centers of power and power alliances in the world, and how their strategies unfold in the short, medium and longer terms. Some of the principal, repetitive cognitive subjects of propaganda are almost as old as the human species itself, representing titanic and enduring struggles over the inequalities and insufficiencies of power, resources, symbolic constructions and belief systems.

In this study I want to establish a middle ground between the principal manifestations, lessons and laws of propaganda, on the one hand, and the adumbration of meta-propagandas on the other. One such meta-propaganda of our time, I believe, is that of neo-conservatism, a propaganda that is most closely associated with the imperial, corporate interests of the USA and its major allies, but which is recognizable in many ways in at least some phases of earlier imperial systems, and has its roots in the battle for domination over scarce resources, and for intellectual and affective hegemony over the hearts and minds of imperial subjects. To understand how such a meta-propaganda is performed and sustained it is necessary to look at specific instances of struggle, not so much in terms of the "manifestations, lessons and laws" of abstracted scientific analysis but in terms of how any particular struggle is articulated within its given context, and why.

# 2 Immediate causes of the 2014 crisis

Ukraine had a population, according to its 2001 census, of approximately 48 million people (down to 45 million in 2014, according to the World Bank [2016]). Of these, some 78 percent were ethnic Ukrainians, 17 percent were ethnic Russians and 6 percent were descendants of other nationalities, including Belorussians, Tatars, Romanians, Lithuanians, Poles and others. Ethnic Russians were distributed unevenly. They were concentrated in the eastern and southern regions where they comprised 58% of Crimea, 39 percent of Luhansk, 38 percent of Donetsk, 26 percent of Kharkov, 25 percent of Zaporozhe and 21 percent of Odessa. In the cities of these regions, the proportion of ethnic Russians was higher (Sakwa 2015: 10–11).

The numbers of Russian speakers was much larger than the identifiable group of ethnic Russians (Sakwa 2015: 59). In terms of language use, 68 percent stated that Ukrainian was their native language, while 30 percent said it was Russian and 3 percent named other languages. Russian speakers were concentrated in the southeast, accounting for as many 91 percent in Sebastopol, 77 percent in Crimea, 75 percent in the Donetsk region and 69 percent in the Luhansk region. Even Kiev was a preponderantly Russian-speaking city. Surveys suggested that these figures underestimated the proportion of Russian-speakers. Citing from Petro, Sakwa referenced the finding of a 2012 study that over 60 percent of newspapers, 83 percent of journals, 87 percent of books and 72 percent of television programs in Ukraine were in Russian, a trend reinforced by the Internet (Sakwa 2015: 59).

While there has long been evidence of a sense among Ukrainians of belonging to a Ukrainian community, even in the Donbass, there have also been and continue to be many and intense historic and cultural links with Russia, such links being perceived as predominantly favorable in Eastern Ukraine but often as unfavorable in the West. Practically every recent election has confirmed that Ukraine is essentially split between those areas in which there is an absolute majority of the Russian-speaking population and those areas in which there is an absolute majority of the Ukrainian-speaking population. The Russian element does not have an influential political party of its own, however, but has been represented through surrogates such as the Communist Party and the Party of Regions (Pogrebinskiy 2015).

Western Ukraine has looked upon ethnic Russians as invaders and occupiers and has nurtured a hatred of Russia on account of Stalin's collectivization campaign in 1932–1933, which contributed to famine and the deaths of two million in Ukraine (often described in Ukraine as the Holodomor – "death-by-hunger" genocide). In 2014, all of Ukraine except Crimea, Sakwa notes (Sakwa 2015: 50), was committed to the idea of Ukraine as a separate sovereign state. The strength of the ethnic Russian presence in Eastern Ukraine and in Odessa likely influenced the vote in favor of Viktor Yanukovych for president in 2010, where voters overwhelmingly preferred Yanukovych over Yulia Tymoshenko (known for her strong anti-Russian views) in Crimea and Eastern Ukraine; Yanukovych won 75 percent of the votes in Odessa.

The immediate cause of the crisis in February 2014 was the earlier decision by President Yanukovych (November 21, 2013) to suspend preparations for signing the "Ukraine–European Union Association Agreement" in Vilnius. This would have opened the borders to goods and eased travel restrictions. Yanukovych argued that Ukraine could not afford to sacrifice trade with Russia, which opposed the deal. He rejected EU/IMF loan packages and their corresponding austerity conditions. A competing deal with Russia would have conceded a $15 billion loan and a reduction of one-third on the price of natural gas for Ukraine. Russia's Vladimir Putin directly charged the USA with instigating the February putsch, a claim he made in the documentary *Crimea – The Way Home*, aired by Rossiya 1 news channel: "They helped train the nationalists, their armed groups, in Western Ukraine, in Poland and to some extent in Lithuania" (RT 2015h). It is important to acknowledge that the unrest in Ukraine 2013–2014 would hardly have been possible were it not for genuine grass-roots discontent with the inefficacy of democratic process and with State and corporate corruption. Sakwa refers to the outcome of the 2004 Orange Revolution as an "oligarch democracy," (Sakwa 2015: 52–73) a brief interval within two decades of endless oligarch wars and self-enrichment of the elite, accompanied by declining living standards and the onset of "stealth authoritarianism." Wilson concludes that under President Leonid Kuchma (1994–2005) Ukraine "was really run by a cabal of oligarchs and regional bosses, in which the president was the chief arbiter" (Wilson 2015: 102), who could draw on the power of "administrative resources" such as State patronage and a legal and tax system to reward friends and punish enemies. By the end of 2015, little seemed to have changed. Wilson attributed poor governance of Ukraine to the continuing influence of former Communists (banned from standing in local elections in August 2015), co-option of opposition, the historical "brain drain" and insufficient energy resources.

The ineptness of the State's response to the Maidan protests was partly responsible for the speed with which the leadership of the Maidan movement was seized by paramilitary elements. These were associated with extreme nationalist, right-wing parties enjoying broad support from many Ukraine oligarchs (whose wings had yet to be clipped in Ukraine, as Putin had done with their Russian counterparts in the 2000s), the USA and its Western allies, the IMF and the European Union. Sakwa comments that "despite its ideological and social heterogeneity, the predominant spirit undeniably was an extreme form of monist Ukrainian nationalism," (Sakwa 2015: 86).

The coverage by the *New York Times* and many other Western newspapers immediately blamed Yanukovych's decision to abort the Ukraine–European Union Association Agreement on "pressure" placed on Yanukovych by the Russian President Vladimir Putin rather than crediting Russia with offering a much better deal. At least equally to blame was Yanukovych's likely realization that an association with the EU would require the acceptance of a loan from the International Monetary Fund (IMF) and conformity with an IMF austerity package that could prove deeply unpopular and damaging to the people of Ukraine, which is indeed what occurred. Russia was offering to put more money on the table, with relatively few conditions. Russia did warn that if Ukraine pulled away from Russia this would have deleterious consequences for Ukraine in terms of the price of Russian gas that Russia's Gazprom supplied to Ukraine, and for Ukrainian trade with Russia (Uhler 2014a, 2014b). These were hardly surprising or ignoble negotiation tactics. Ignoring these and many other nuances, the narrative told by Western media constructed a misleading image of a popular "revolution" against Yanukovych's corrupt, pro-Russia government rather than of a coup propelled beyond its initial grass-roots proponents by fascist elements, supported by the USA, and who at best represented only a modest fraction of the political spectrum of northern Ukraine. Even George Friedman, the Founder and CEO of Stratfor – sometimes described as the "Shadow CIA" firm – declared in a December 19 interview for the Russian magazine *Kommersant*, that the overthrow of Ukraine's President Viktor Yanukovych "was the most blatant coup in history" (Zuesse 2014d). Yanukovych, after all, had come first place in the January 2010 election with 49 percent of the vote, an election that international observers had considered relatively free and fair (Sakwa 2015: 56).

Western reporting of the riots that followed Yanukovych's decision focused on the presumption of pressure from Moscow as the reason for his turnaround from Europe; they downplayed the IMF austerity conditions, overlooked the interests of Eastern and Southern Ukrainians – many of them Russian-speaking ethnic Russians who did not support the objectives of the protestors in Kiev – and they failed to factor into the equation the fact that Yanukovych's ruling party had legitimately won the most recent general election, was already due to present itself to the electorate within months of the protests, and had managed to win four of five parliamentary seats during by-elections in this period. Protestors, now spearheaded by the extreme Right, ignored the terms of an agreement brokered between the Opposition and the Ukrainian government by the EU (represented by France, Germany and Poland). This would have accelerated elections and established a national unity government. At precisely this point, the demonstrations turned into an uprising. The EU itself took no measures to ensure that the Opposition respected the agreement that they reached. Yanukovych agreed to speed up the 2014 elections and ordered police to stand down. At that point the neo-Nazi Right Sektor and similar storm troopers occupied government buildings. Yanukovych and his aides fled.

Once the protestors had succeeded in forcing out Yanukovych and many of his administration's leadership by threat and intimidation, they sought to legitimize their takeover by converting the Ukrainian Parliament (Verkhovna Rada) into a rubber stamp legislature. Even while half of the parliamentarians were absent or in hiding, the coup regime leveraged the very instability they had provoked to declare its unchallenged Rada bills as legitimate (Nazemroaya 2014a). The country's biggest party, the Party of Regions, and other pro-government parties or independent parliamentarians were not present for all the Rada votes. There were many departures from protocol, including threats by the coup leaders to ban the Party of Regions, holder of almost 40 percent of the seats in the Rada (Nazemroaya 2014a). The Party of Regions disassociated itself from Yanukovych following his departure. Many deputies left the Party to form other groups, and many were later pursued with criminal charges. The Party did not participate in the September parliamentary elections. By that time the number of its deputies had fallen from a 2012 high of 210 to a low of 80.

Yanukovych and the Party of Regions had initially supported the Ukraine–European Union Association Agreement (a process which dragged out over three years, with its text not even available in Ukrainian until the last moment), but they backed out when the EU refused to renegotiate the agreement or to give financial guarantees and economic relief to Kiev to help compensate for the trade losses and higher gas prices that Kiev would likely face if it turned its back on Russia. Some Ukrainian oligarchs worried that the agreement would allow EU corporations to dismantle Ukrainian corporations and change many of the trade laws and regulations that advantaged them. Yanukovych, says Sakwa, realized rather late in the day that the EU had put very little on the table that would be of immediate benefit while threatening destabilization. Sakwa describes the agreement as 100,000 pages of law and regulations which were economically exclusionary towards Russia, even though Russia was Ukraine's largest trade partner accounting for 24 percent of its exports and 30 percent of its imports (Sakwa 2015: 75). Russia repeatedly warned that once the Ukrainian market was open to EU goods, the Russian market would be flooded with lower-quality Ukrainian items, and that the agreement could provoke social unrest and the possible secession of pro-Russian regions.

Western media generally reported the events of what they liked to call Euro-Maidan as: (1) a positive development in support of democracy, transparency and independence; (2) representative of the people of the Ukraine as a whole; or, (3) if not exactly representative, then good for them. Why good? Summarizing the argument of the scholar Halyna Mokrushyna (2015c), the answer was: because the West is good; neoliberalism is good; Ukrainians who do not want to be part of the West or part of neoliberalism must be stuck in a Soviet past that is totalitarian, bureaucratic and dated. This viewpoint was deaf to the actual diversity of positions within Ukraine. To paint Yanukovych merely as a puppet of Russia was to gravely underestimate the internal dynamics of power in an overly centralized, ethnically heterogeneous State. WMM gave insufficient weight to how many Ukrainians, especially in the East, entertained positive sentiments for the former Soviet Union as a collectivist, socialistic, idealistic entity that had struggled to provide full employment, free education, free healthcare, free or low-cost housing and social security. Deeper even than the Soviet memory, there was also a Byzantine orientalist tradition in Eastern Ukraine. Mercouris cites an October 2014 survey, conducted jointly by the Kyiv International Institute of Sociology and Ilko Kucheriv Democratic Initiatives Foundation. This found that less than 20 percent of the population of Ukraine took part in the protest actions of November 2013–February 2014 in Kyiv (EuroMaidan). The biggest participation occurred in Western Ukraine, where roughly half of the population was involved in different ways, but elsewhere the situation was very different:

> In the central regions, participation was much lower: 9.5% took part in protests in Kyiv and 2% in other cities and villages. The participation of Ukrainians from other regions was significantly lower: 2% in the southern regions; 3% in the eastern regions, and 3% in Donbas. The data of non-participation is as follows: western regions – 46.9%; central regions – 80.9%; southern regions – 96.6%; eastern regions – 95.1%; Donbas region – 97.1%. The national level non-participation of Ukrainians in Euromaidan was 81.9%. Yet, after their Euromaidan revolution won (as the officials and media in Ukraine claim), one fifth of "revolutionary" Ukraine started to reform the remaining four fifths.
>
> (Mercouris 2015b)

There was much sound cause for popular unhappiness with the Yanukovych government in 2013. The mistake of many WMM was to overlook an enduring context of popular discontent whose fundamental causes were political centralization, a lackluster economy, inequality, the overweening power of oligarchs, and commercial and political corruption. These were at least as much in evidence in 2015 as they were in 2013. Western media rushed to cast EuroMaidan in the same positive light as they had welcomed the manifestations of the Arab Spring across 2010–2012 – regime-changes whose benefits were already highly questionable – and to misrepresent Russia as an antediluvian, authoritarian villain engaged in aggressive battle against a virtuous, democratic, freedom-loving and peace-loving EU, NATO and US alliance. Such broad-brush stereotyping made it very difficult for WMM to then pay rigorous attention to how this same alliance manipulated the situation to its own advantage, and created an opening for extreme nationalist parties.

# 3 Liberal freedom fighters or fascist militia?

## Legitimating a coup regime

A *Washington Post* headline "Ukraine Decries Russian 'Invasion'," typified mainstream Western press's demonization of Russia, treating a coup regime as though it spoke for the entire country, whereas in reality it spoke for only a subset of the population, mostly from Western Ukraine (Parry 2014b). The media followed the lead of Washington whose administration was keen to legitimize the illegitimate. Indeed, Washington was party to the illegitimacy, having openly encouraged the protests that led to the coup. The tactics of the Western legitimization campaign included: (1) attempts to blank out or downplay the presence of fascists among leading parties to the coup; (2) staging, at unseemly speed, the coup regime's Prime Minister Arseniy Yatsenyuk's formal welcome at the White House as early as March 12, 2014; (3) pretending that the concerns of ethnic Russians in Ukraine's east and south were merely the outcomes of Russian propaganda and manipulation; (4) maintaining that calls for secession of Crimea from Ukraine were illegitimate (even while bolstering the pretense of the legitimacy of the coup regime in Kiev); and (5) pompously calling for and enacting sanctions targeted against specific Russian leaders as punishment for their supposed malfeasance.

While WMM more or less uniformly deplored Russia's "invasion" of the Crimea, they often seemed incognizant of the fact that the alleged invasion involved little by way of military forces crossing an international border (Parry 2014m). Rather, Russian troops were already stationed in Crimea, legitimately, under a long-standing agreement with Ukraine's government.

> The impetus for Crimea's secession from Ukraine and rejoining Russia came from the local government and the Crimean people, not from Russian military force. But repetition of the words "invasion" and "annexation" is needed to elicit the desired revulsion from the American people.
>
> (Parry 2014m)

A Russian source interviewed by Whitney (2014i) claimed that additional troops were indeed sent to the Crimea by Russia – at a moment when it seemed as though Kiev would launch an attack on Crimea, and in the context of the legal vacuum that followed the forced departure of democratically elected Yanukovych – but that the total numbers never exceeded the quota of Russian soldiers legally permitted in the Crimea.

WMM narratives of the 2014 Ukraine coup were deeply biased. Parry (2014e) wrote of the *New York Times* that virtually everything the paper had to say about the Ukraine was:

> so polluted with propaganda that it requires a very strong filter, along with additives from more independent news sources, to get anything approaching an accurate understanding of events. Since the early days of the coup, the Times has behaved as essentially a propaganda organ for the new regime in Kiev and the State Department, blaming Russia and Putin for the crisis.
>
> (Parry 2014e)

Walter Uhler echoed Parry: "After conducting an extensive investigation of the *Times*' reporting on events in Ukraine, I have reached the same conclusion. *The Times* has become an irresponsible propaganda bullhorn" (Uhler 2014b). WMM support for the coup regime in Kiev was widespread. There were some indications that popular sentiment was strongly opposed to WMM's "normalization" of the illegitimate. Rombach (2014) provided a detailed account of mainstream media support for Kiev in Germany but also recorded a very high level of reader and audience discontent with mainstream coverage, as illustrated in audience surveys and readers' letters and comments.

Part of the legitimizing process encompassed the attribution of anti-Kiev views, as in Eastern Ukraine, to the success of "Russian propaganda." Vera Graziadei undertook a critical analysis of a single piece by the BBC Diplomatic Correspondent, Bridget Kendall. Kendall's article was headlined with the sentence: "Russian propaganda machine 'worse than Soviet Union,'" indicating that it had gripped the whole of Russia at the time of her visit to Moscow in June 2014. Graziadei identified what she claimed were thirteen propaganda techniques in thirteen paragraphs of Kendall's article, including the use, presumably as "evidence," of Russian "celebration" of its "seizure" of Crimea – a photo of two people dancing that had in fact been taken four years previously (Graziadei 2014). What followed later in terms of WMM coverage, which I shall examine from a number of different perspectives, continued to function as legitimation of a coup that was not merely illegitimate, but was being used as a pretext for significant destabilization of Russia's economy and, more significantly, as a false pretext for aggression against an armed and nuclear power.

## Role of right-wing militia and oligarchs behind the EuroMaidan

Triggering the 2014 ouster of Yanukovych was the latter's decision not to sign the Association Agreement with the EU. On November 21, 2013, Mustafa Nayyem, an Afghan-born journalist for an online news site, used Facebook to call on people to gather on the Maidan, a central square located on Khreshchatyk Street in the Shevchenko Raion in Kiev, to protest. At the beginning, says Sakwa, the protesters were mainly middle-aged and middle-class, looking to the EU to escape economic and political stagnation (Sakwa 2015: 82). Certainly, Ukrainian police repression of riots on November 29–30, 2013 was brutal, but was followed on December 1 by a violent storm of hundreds of protestors against the unarmed police line near the presidential administration building, in turn followed by a counter-attack from riot police (Uhler 2014b, citing Deputy Director of the Center for Society Research [Kiev], Volodymyr Ishchenko). It was abundantly clear, even at this early stage, that the chants of the opposition parties included Nazi-sympathetic slogans, that members of the Right Movement (Right Sektor) had undertaken training for violent resistance to law enforcement, and that opposition weaponry included stun grenades. The four main opposition parties had formed an alliance with Arseniy Yatsenyuk as leader.

Maidan protestors had initially demanded that the Ukrainian government simply reverse its decision not to sign the Ukraine–European Union Association Agreement (Roberts 2014a). Ultra-nationalists introduced violence into the protests and elevated their demands from signing with the EU to overthrowing the elected government (even though legitimate elections were due within months). In Lviv, protesters occupied several government buildings, festooned the City Hall with Nazi banners and dispatched several hundred armed forces daily to the Maidan. The incumbent regime made many mistakes, not least the bundle of draconian "anti-protest" laws adopted for immediate application and in contravention of its own rules, by the Verkhovna Rada on January 16, 2014. In the counter-narrative of many alternative media, overreaction of the incumbent authorities allowed "Western stooges" to seize power, disband the police, issue arrest warrants for the president, and intimidate parliament, public prosecutors and television editors.

The *New York Times* paid little attention to the Right Sektor's role in the protests, acknowledging Svoboda on December 16, 2013. It picked up on Right Sektor on February 1, 2014 but paid it no serious attention until mid-February (Uhler 2014a, 2014b). The *New York Times* correspondents, Uhler complained, acknowledged the role of Svoboda and its Nazi links while tending to minimize the violence. Reports of the riots that had appeared on December 1 said nothing about Right Sektor violence even though, Uhler claims, "numerous investigations show that the overwhelming majority of attackers were the far right and neo-Nazi militants from the so-called 'Right-wing sector,' which unites various nationalist groups participating in Euromaidan" (Uhler 2014b). Worse, found Uhler, *The Times*: "failed to characterize, let alone highlight, any behavior by protesters as crimes – even after the Shevchenko district court in Kiev ruled to prohibit the blocking of government buildings and the obstruction of government activity" (Uhler 2014b).

In a phone discussion leaked to YouTube early in March, the Estonian Foreign Minister Urmas Paet told the EU Foreign Affairs chief Catherine Ashton that people on both sides of the conflict – protesters and police – had been killed by snipers during the Maidan Square protests. Snipers had been hired by leaders of the opposition and not, as reported at the time, by the government. Western media that only a few days previously were attributing shootings to the authorities were now inexcusably slow to pick up on these revelations. Five days after the leak, Media Lens declared itself unable to find more than one reference to the YouTube leak on the BBC website, and even that one was buried in an article about something else. The *Guardian* covered the story, now confirmed by the Estonian Ministry, but described it as a "conspiracy theory." By March 10, there had appeared only seven mentions of the leaked phone call, by comparison with 250 reports on snipers prior to the leak, of which many had attributed the shootings to the authorities in power. David Peterson (2014), conducting a NewsBank search covering 3,253 sources for March 3–7, found only a total of eight items in the US-based media that mentioned the Paet revelation.

The Paet revelation was endorsed by a report on the German public television show *Monitor* in April. Evidence suggested that radical oppositionists deliberately involved professional snipers as a way to create chaos and place the blame on Yanukovych. This was indicated by the shootings on February 20 that came from the rooftop of the opposition-controlled Hotel Ukrayina and that fired into the backs of protestors as they marched towards government buildings (Stern 2014). One commentator noted that the role of snipers: "appears to be a staple feature of the US-arranged revolutions. Snipers killing both protesters and police were reported in Moscow's 1991 and 1993 revolutions, as well as in many other cases" (Shamir 2014).

Further investigation and prosecution of those criminally responsible was highly unlikely. The Attorney-General of the coup government, Oleg Machnitzki, was a representative of the right-wing nationalist Svoboda party. A European Council report in March 2015 judged that the attitude towards investigations into Maidan violence manifested by Ukraine's Interior Ministry was "uncooperative and obstructive," and that appointment, post-Maidan, of certain officials to senior positions in the Ministry "contributed to the lack of appearance of independence" and "served to undermine public confidence in the readiness" of the Ministry to investigate. Whatever investigations were attempted at the time were insufficiently coherent or effective (RT 2015c).

In Kiev, official investigations appeared to have stalled in acrimony at the end of 2015, almost two years after the Maidan events and amidst a high level of resentment towards the authorities on the part of families of victims. Preliminary findings claimed that seventy-seven protestors had been killed during the ninety-five days of unrest. An additional 185 were wounded by gunfire, and another 1,000 suffered other wounds. Some activists considered that the actual numbers were a great deal higher. According to Vitaly Tytych, a lawyer representing some victims' families, the authorities had planned a crackdown on the protests of February 18, using live ammunition and employing irregular killing squads. A total of 274 officials were under investigation, including even senior Interior Ministry officers but none had been suspended. Tytych considered that investigators were coming under pressure by the "old guard" in the Prosecutor

General's office (Andersen 2015b). The official investigation, stalled as it was, and as reported by a regime-friendly source, the *Kiev Post* (Western media no longer showing much, if any, interest) appeared to be principally focused on the wrongdoings of the Yanukovych regime. Meanwhile, an academic study (The "Snipers' Massacre" on the Maidan in Ukraine) by the political science researcher Ivan Katchanovski (2015b) was cited by *Sputnik News* in January 2016 as concluding that the killings of protestors were organized as a false flag operation by far-right paramilitary groups and allied political parties including Right Sektor, Svoboda and Fatherland – and not by the former government's Berkut riot police. The report drew on analysis of eyewitness reports, ballistic trajectories, security forces' radio intercepts, 5,000 photos, 1,500 videos and broadcast recordings. The study cited numerous media manipulations of evidence to support the dominant but false Kiev narrative of what had happened. In its conclusion, Katchanovski stated:

> The study shows that concealed armed Maidan groups, which were based in particular in Music Conservatory and the Trade Union buildings, started the massacre in the early morning on February 20 by targeting Berkut and Internal Troops units on the Maidan itself with live ammunition fire, inflicting their mass casualties, and forcing them to retreat. Then the armed Maidan groups, in particular the same ones, massacred the unsuspecting Maidan protestors from concealed positions. The analysis of circumstances, timing, and locations of specific killings of 49 protestors presents evidence that almost all of them were killed from the Maidan-controlled buildings and locations, in particular the Hotel Ukraine and Zhovtnevyi Palace. This study presents direct evidence, such as videos, photos, and witness testimonies, concerning groups of the Maidan snipers in these buildings and their shooting from these positions in the directions of the protesters at the same time when the protesters were killed and wounded from these directions from the same caliber and types of weapons.
>
> (Katchanovski 2015b)

In a May 2014 conference, President Obama attempted to justify US and NATO support for the coup regime with reference to Article 5 of the NATO Treaty. This states that: "The Parties agree that an armed attack against one or more of them in Europe or North America shall be considered an attack against them all." Had the President conceivably overlooked the facts that: (1) Ukraine was not (yet) a member of NATO; (2) that the regime which Obama was pledging to support had come to power via a coup; (3) thus establishing a regime in which the vice-premier, minister of defense, minister of agriculture and minister of environment all belonged to far-right neo-Nazi parties (Blum 2014)? Even less remarked by either mainstream or alternative media during the early stages of the crisis was the role of EU members, in particular of Germany. When the German premier visited Kiev several months further into the crisis, she expressed enthusiasm for Ukraine's promise as a low-wage center and cheap supplier of raw materials (Dreier 2014).

Not only was the role of fascists in instigating the coup routinely overlooked by WMM, but also the role of oligarchs – some of them, paradoxically, Jewish – in exploiting the conditions of chaos that they had helped finance. The most generous supporter of right-wing militia was multi-billionaire Ihor "Benya" Kolomoiski, whose immediate reward from the coup regime was the governorship of the Russian-speaking city of Dnepropetrovsk:

> He was not required to give up his Israeli passport. His brethren oligarchs took other Russian-speaking industrial cities, including Kharkov and Donetsk, the Ukrainian Chicago or Liverpool. Kolomoiski is not just an "oligarch of Jewish origin": he is an active member of the Jewish community, a supporter of Israel and a donor of many synagogues, one of them the biggest in Europe. He had no problem supporting the neo-Nazis, even those whose entry to the US had been banned because of their declared anti-Semitism. That is why the appeals to Jewish consciousness against the Brown putsch demonstrably failed.
>
> (Shamir 2014)

The coup rule imposed harsh conditions for ethnic Russians and Russian speakers of Eastern and Southern Ukraine. Since 2005 the Russian language had already suffered a full frontal attack in education and the media. The coup regime banned the Communist Party and suppressed the Party of the Regions (the biggest in the country). Its first decree banned the Russian language from schools, radio and television, and forbade all official use of Russian, restoring Ukrainian as the sole state language at all levels. This measure, while it was not signed into law, was correctly interpreted by many ethnic Russians and pro-Russian speakers as an indication of implacable hostility towards all things Russian. There were proposals to arrest Ukrainians who held dual Russian/Ukrainian citizenship. War memorials to the Red Army – that had liberated Ukraine from Germany in World War II – and even memorials to the Russian general who had defeated Napoleon, were destroyed (Roberts 2014b). The banned Russian television channels included Vesti, Russia-24, Channel One International, RTS "Planeta," and NTV World (Lendman 2014). Shamir reported that:

> The Minister of Culture called Russian-speakers "imbeciles" and proposed to jail them for using the banned tongue in public places. Another decree threatened every holder of dual Russian/Ukrainian nationality with a ten-year jail sentence unless he or she surrendered their Russian nationality right away.
>
> (Shamir 2014)

Such measures were backed by violence. Storm troopers of the coup regime:

> went around the country terrorizing officials, taking over government buildings, beating up citizens, destroying Lenin's statues, smashing memorials of

> the Second World War and otherwise enforcing their rule. [...] They began to hunt down riot policemen who supported the ex-president, and they burned down a synagogue or two. They tortured a governor, and lynched some technicians they found in the former ruling party's headquarters. They started to take over the Orthodox churches of the Russian rite, intending to transfer them to their own Greek-Catholic Church.
>
> (Shamir 2014)

In his book, *Old Nazis, the New Right, and the Republican Party*, Russ Bellant exposed the deep involvement of the US Republican Party in Nazi elements from Central and Eastern Europe, including Ukrainians. These dated back to World War II and even earlier (Bellant 1991). Bellant told Rosenberg (2014) that a key organization in the 2014 Kiev coup was the Organization of Ukrainian Nationalists (OUN), or a specific branch of it known as the Banderas (OUN-B), the group behind the Svoboda party. The Svoboda party had been condemned by the World Jewish Congress and by the EU Parliament as recently as December 13, 2012. Viktor Yanukovych, deposed in February 2014, had officially annulled his predecessor's award of the title "Hero of Ukraine" to Bandera. In April 2015, the Kiev parliament voted to extend official recognition to Bandera's Ukrainian Insurgent Army, a measure that provoked the wrath of the Simon Wiesenthal Center, which declared that Bandera's Army had participated in the mass murder of Jews. The Center also condemned a simultaneous measure to equate Communist with Nazi crimes (Sokol 2015).

The OUN split off from other groups in the 1920s to throw off Polish rule, and from the 1930s launched a campaign of assassinating and otherwise terrorizing people who didn't agree with them. The OUN-B was a splinter group formed in 1940 under the leadership of Stepan Bandera and backed, during World War II, by a Gestapo-supported secret police official, Mykola Lebed. The CIA was later to help Lebed gain entry to the USA, along with many of his kind who were seen as assets for the organization of anti-Soviet resistance. The OUN-B also collaborated closely with the German foreign intelligence service Abwehr. Thus it allied with German Nazis before and during World War II and assisted in the German invasion of the Soviet Union. Members participated in the identification and rounding up of Jews in Lviv, assisting in or executing several thousand citizens. They also participated in liquidating Polish groups in other parts of Ukraine during the war. They were backers of the formation of the 14th Waffen SS Division, an all-Ukrainian division that became an armed element on behalf of the Germans and operating under overall German control. Alongside them in the Waffen SS were Hungarian and Latvian Nazis. All of them were involved in the mass murder of Jewish citizens and of co-nationalists who were anti-Nazi (Marshall 2016).

Bandera himself was imprisoned by the Germans. During this time, he organized the Ukrainian Insurgent Army (UPA). He was released in 1944 when Germany supported his efforts to organize resistance to the Soviets. After the war, Soviet forces killed, arrested or deported several hundred thousand members, relatives or supporters of the UPA and OUN. Bandera was assassinated by the KGB in 1959.

To this day, the city of Lviv is still host to extreme Ukrainian nationalism. It supplied 600 protesters a day to the Maidan protests of February 2014 (Marshall 2016). The OUN continued to glorify the history of the Waffen SS after World War II. SS division insignia were worn by some of the marching protestors during the Maidan protests of 2014. The OUN had called for the establishment of ethno-genetically pure Ukrainian territory, an aspiration also expressed by leaders of Svoboda in 2014. The USA had maintained long-standing ties to the OUN through its intelligence agencies.

After World War II, the leadership of organizations like the OUN and comparable organizations in the USA (involving up to 10,000 members) that had fought for the Germans in other parts of the Soviet Union organized "captive nations" committees, and came to be regarded as representatives of the oppressed peoples of Eastern Europe and the Baltic countries. But what they represented was the most extreme elements of each of these national communities. The committees became politically active through the Republican National Committee following the policy decision of the Eisenhower administration in the early 1950s to consult with them. Richard Nixon essentially unified these "Captive Nations Committees" on a permanent basis as the ethnic outreach arm of the Republican National Committee, and helped to rehabilitate Laszlo Pasztor, who had served five years in prison after World War II for crimes against humanity.

One of the movement's 1984 booklets praised the pro-Nazi Ustashi regime of Croatia. Ustashi had killed an estimated 750,000 people, burning them alive in their own camp in Croatia. With sponsorship from the American intelligence agencies, OUN members became involved in Europe in a variety of ways, as in contributing to the CIA's Radio Free Europe headquartered in Munich and running operations in the Soviet Union. Some members moved back into Ukraine and other post-Soviet territories and began setting up operations there. They reconstituted the veterans group of the Waffen SS, held marches in 1990s Ukraine, and organized political parties with assistance from the USA. OUN members were active in what came to be called the Orange Revolution in 2004 when they helped win the election. Viktor Yushchenko, president of Ukraine from 2005 to 2010, was their close ally. They worked with the new government to get benefits for Ukrainian SS division veterans, and started establishing the statues, memorials and museums for the OUN founder and leader, Stepan Bandera (Marshall 2016).

In a critique of Western leaders and media, the investigative journalist John Pilger observed that:

> for the first time since 1945, a neo-Nazi, openly anti-Semitic party controls key areas of state power in a European capital. No Western European leader has condemned this revival of fascism in the borderland through which Hitler's invading Nazis took millions of Russian lives. They were supported by the Ukrainian Insurgent Army (UPA), responsible for the massacre of Jews and Russians they called "vermin." The UPA is the historical inspiration of

> the present-day Svoboda Party and its fellow-travelling Right Sektor. Svoboda leader Oleh Tyahnybok has called for a purge of the "Moscow-Jewish mafia" and "other scum," including gays, feminists and those on the political left.
>
> (Pilger 2014)

WMM tried to overlook or discount the role of Nazis in the coup against the Yanukovych government. Media reporting operated on the assumption that any movement that wanted to affiliate with Europe and not with Russia must be a force for good – failing to appreciate that, for many of the protestors, Europe stood not for democracy and enlightenment but represented a bulwark against Russia that in turn would facilitate a vigorous and in many ways retrograde, even fascist nationalism. Right Sektor and Svoboda party activists decked out some of the occupied government buildings with Nazi insignias and even a Confederate battle flag, the universal symbol of white supremacy. Until 2004, Svoboda was still called the Social-National Party of Ukraine employing neo-Nazi and SS symbols. Even after it changed its name and symbols, its leader, Oleh Tyahnybok, railed against the "Muscovite-Jewish mafia running Ukraine" and praised the Ukrainian Insurgency Army (UPA) in World War II for fighting "against the Moscali [Muskovites], Germans, Zhydy [Jews] and other scum, who wanted to take away our Ukrainian state" (Uhler 2014a, 2014b).

The veteran investigative journalist Robert Parry was more assiduous than most in following this narrative for his alternative online site, *ConsortiumNews.com*. Parry had also noted that Stepan Bandera was one of the heroes celebrated during the Maidan protests. The historian Richard Sakwa observed that a giant portrait of Bandera was positioned to one side of the stage during the Maidan protests (Sakwa 2015: 19). At the height of the violence in Kiev 2014, February 20–22, neo-Nazis led the movement with 100-man brigades called "the hundreds," launched assaults against police, and forced Yanukovych and many of his officials to flee for their lives. In the regime that emerged from this chaos, with the assistance of Western governments, four ministries were awarded to right-wing extremists. These included the Ministry of National Security which was gifted to Andriy Parubiy, founder of the Social-National Party of Ukraine in 1991 which blended Ukrainian nationalism with neo-Nazi symbols. Prior to his post-coup elevation, Parubiy was commandant of the Maidan's "self-defense forces" (Parry 2014q).

Fascism was again in evidence in Odessa, early in the ensuing May, when right-wing youths attacked an encampment of ethnic Russian protesters, driving them into the Trade Unions Building. This was then set on fire with Molotov cocktails. A total of forty-eight people were killed. Dozens died inside the building as their tormentors sang the Ukrainian national anthem. People who tried to flee were chased and beaten. The building was spray-painted with swastika-like symbols and graffiti reading "Galician SS," a reference to the Ukrainian nationalist army that fought alongside the German Nazi SS in World War II, killing Russians on the Eastern Front (Parry 2014l).

The Odessa incident, which some consider signified the beginning of a real "civil war" in Ukraine, was a great deal more shocking than anything first reported. Many of the anti-junta activists who occupied the Trade Unions House, wrote Whitney, were:

> savagely shot at point-blank range by agents and thugs who had infiltrated the building to kill as many of the occupants as possible, burn the corpses, and then slip away without notice. Some of the victims – like a young woman who was eight months pregnant – were strangled with an electrical cord.
>
> (Whitney 2014a)

In another case, a woman was stripped naked from the waist down, raped, killed and set ablaze. Some victims had flammable fluid dumped on their heads and were incinerated, leaving a charred head atop a corpse whose clothes were untouched by fire. Whitney considered that these details pointed towards a carefully planned black-op, one that likely involved foreign intelligence agencies working hand-in-hand with the fascist junta government in Kiev.

In covering the atrocity, Western media protected the Kiev coup regime by minimizing the horror or, following the example of the *Wall Street Journal*, even cited Kiev regime sources in order to pin the blame on the ethnic Russian "rebels" themselves for burning down the building. This allegation flew in the face of the widely available video footage of neo-Nazis throwing Molotov cocktails at the building. Whitney referred to the video footage shown by *Russia Today* that exposed how agents in red armbands, mingled in amongst ethnic Russian activists, had initiated a confrontation with the police and then quickly switched sides when the fighting broke out – a classic false flag operation. The police had immediately opened their ranks to let the imposters slip by when the street-scrum began. The same agents were later filmed shooting handguns and automatic weapons in the direction of the Trade Unions Building, just minutes after they had switched sides (Whitney 2014a).

The Council of Europe, in September 2014, extended the mandate of its International Advisory Group in Ukraine to examine whether the Odessa investigations met the requirements of the European Convention on Human Rights. The International Advisory Group had been established by the secretary-general of the Council of Europe in April 2014 to oversee investigations into violence during the EuroMaidan protest in Kyiv. The European Council report in November 2015 was extremely critical of official investigations of the Odessa events of May 2014. It concluded that substantial progress had not been made. Ukraine government investigations fell short of European standards. Official probes failed to comply with all of the requirements of the European Human Rights Convention. The first forensic report, not prepared until July 2014, did not include an on-site inspection. Another, ordered in April 2015, was still underway the following August. Official investigations into the Odessa events were "ineffective," and authorities had failed "to show sufficient thoroughness and diligence in initiating and pursuing" the inquiries. Until December 2014, there was no real effort to account for the delay of over forty minutes in the arrival of firefighters to the Trade Unions Building. The government's investigation into the street violence and fire, plus a separate inquiry into the conduct of emergency services staff during the fire, "lacked institutional and practical independence." Enquiries carried out by the interior minister and the State Emergency Services should have been carried out by organs entirely independent from the police and fire services. The report expressed "concern" about the prosecution and trials of suspects, particularly about decisions to terminate the proceedings against two suspects on grounds of lack of evidence. The authorities had not taken "coordinated measures" to inform victims and next-of-kin about the progress of the investigations (RFE/RL 2015a).

The US media tried to dismiss evidence of fascism by ascribing it to Russian propaganda. In his April 20 column, the *New York Times* columnist Nicholas Kristof dismissed talk of fascism as nonsense while visiting his ancestral home in the western Ukrainian town of Karapchiv (Kristof 2014). He had underestimated the evidence amassed in a Russian Foreign Ministry "White Book" that chronicled Kiev human rights violations from November 2013 through March 2014. These included violations of human rights, internal interference by the EU and USA, weapons and violent methods used by protestors, restrictions on basic freedoms and crackdown on dissidents, discrimination based on ethnic background and religious persecution (Voice of Russia 2014).

Some Ukrainian media openly ascribed to Nazi sentiment. Zuesse observed how on Hromadske TV:

> with the ethnic cleansing in Donbass raging full force [...] an unknown "journalist," Bogdan Butkevich, was interviewed, and said, "Donbass must be exploited as a [gas-producing] resource, which it is. [...] Donbass [...] is severely overpopulated with people nobody has any use for [...] approximately 4 million inhabitants, at least 1.5 million of which are superfluous. [...] There is a certain category of people that must be exterminated."
>
> (Zuesse 2014b)

During the lead-up to the coup, this station received 8 percent of its funding from Soros's International Renaissance Foundation, 16 percent from the US Embassy, 31 percent from the Netherlands Embassy and 44 percent from "Individuals," including Ukraine's oligarchs such as gas baron Ihor "Benya" Kolomoiski already cited. Other oligarchs acquired political control of Russian-speaking industrial cities, including Kharkov and Donetsk. Zuesse took the *New York Review of Books*' reporter Tim Judah to task for Judah's presumption in an article on April 28 (Judah 2014). Judah maintained that the belief among secessionists that the Kiev coup plotters were Nazis and NATO-backed was so obviously the result of Russian propaganda that it wasn't worth following up. Rather than manufacture stretched analogies with (actually correct) Serbian accusations of Nazi links in Croatia during the break-up of Yugoslavia, Zuesse suggested Judah should read south Ukrainian resident George Eliason's history of the Ukraine in *OpEdNews*. Eliason had recorded Nazi genocide perpetrated by the Banderite (north Ukrainian) movement and Banderite links to the recent coup plotters, some of whom had designs on the "liberation" of regions located in Russia. Yulia Tymoshenko, a candidate for the presidency, had even called for the nuclear obliteration of Ukrainian Russians. Her ally, billionaire Ihor Kolomoiski, Zuesse claimed, had masterminded the May 2, 2014 rounding-up and burning of Odessa's proponents of independence from the Kiev regime, which roasted more than 200 civilians (Eliason 2014).

While welcoming the presidential elections at the end of May, the *New York Times* Editorial Board itself acknowledged that it was a contest between "chocolate king" Petro Poroshenko, the recently released "gas princess" Yulia Tymoshenko, and the banker Sergey Tigipko, all of whom were members of "the clique of very rich businessmen who have been at the root of the corruption of the Ukrainian government" and who "made their fortunes in the chaotic privatization of state assets" following the dissolution of the Soviet Union. In other words, noted Van Auken, the Ukrainian people were "being asked to choose between a trio of criminals and parasites who enriched themselves at their expense through the theft of state property" (Van Auken 2014a). The election turnout greatly over-represented the regions that had favored the coup and underestimated those in the South East that had not, where the vote fell below 50 percent (Zuesse 2014b). The peoples in the South East, in the most recent authentically democratic previous election – February 2010 – had voted overwhelmingly for Viktor Yanukovych. Poroshenko won by an absolute majority in the first round, although he and his opponent Tymoshenko combined received fewer votes than Yanukovych had in 2010 (Sakwa 2015: 123). Sakwa argues that although Poroshenko gained a popular mandate he was a "prisoner of the Maidan," that is to say, he was answerable mainly to extreme nationalist, "monist" conceptions of Ukraine as opposed to pluralist alternatives. Additionally he came under increasing pressure from divisions even within the pro-Western blocs. In one incident towards the end of 2015, a member of Poroshenko's bloc physically picked up Prime Minister Arseniy Yatsenyuk and pulled him from the podium (Reuters 2015l).

Writing from Mariupol in September, the *Guardian* journalist Shaun Walker observed disturbing indications of Nazism among pro-Kiev Azov militia fighters whose battalion received funding and assistance from the governor of Donetsk region, the oligarch Serhiy Taruta (Walker 2014c). Similar observations were later made in June 2015 by William Cathcart and Joseph Epstein (Cathcart and Epstein 2015). They noted that Azov answered directly to the Ministry of Internal Affairs, worked closely with Ukrainian intelligence, SBU, and that the Kiev government depended on Azov for the protection of Mariupol, which Azov had "liberated" in the summer of 2014. In December 2015, up to 5,000 ultranationalists marched through the central avenue of the city alongside the Azov fighters carrying burning torches and Azov flags with the battalion's insignia. Although Walker concluded that fears of Nazism in Kiev were overblown, he considered that the Azov fighters would pose a threat to the Ukrainian government itself, once fighting in the East was over. The battalion's symbol was reminiscent of the Nazi Wolfsangel, many of its members had links with neo-Nazi groups, and it had attracted far-right volunteers from abroad. Several of those who spoke to Walker considered that Ukraine would benefit from dictatorship. The Azov militants led the way in pulling down Lenin statues in cities such as Kharkiv, as observed by the Organization for Security and Co-operation in Europe (OSCE) monitors, and in launching attacks on the Donetsk and Luhansk People's Republics. In December 2014, they blocked the distribution of humanitarian aid to rebel-controlled areas in Eastern Ukraine. James Carden struck a parallel note of alarm the following January, citing warnings from Atlantic Council Senior Fellow, Adrian Karatnycky, that far-right battalions like Azov and Dnepr-1 were threatening to take over the Kiev government if Poroshenko failed to defeat Russia (Carden 2015). A further danger was that of US and other Western funding supporting such groups. Carden cited an interview with journalists Will Cathcart and Joseph Epstein in which a member of the Azov battalion spoke about "his battalion's experience with U.S. trainers and U.S. volunteers quite fondly, even mentioning U.S. volunteers engineers, and medics that are still currently assisting them" (Carden 2016).

Battalions also drew strength from Islamist jihadists, some of them ISIS linked, using Ukraine as a staging post where they could buy Ukrainian citizenship, passports and weapons before seeking entry to Europe or linking up with Ukrainian volunteers who had supported them in the Chechen wars (Mamon 2015b; Nazemroaya 2015). The Jihadi connection was acknowledged by the *New York Times* in July 2015, in a report of several volunteer Islamic battalions, mainly from Chechnya, fighting on the Ukrainian side (Kramer 2015b). One of the goals of foreign fighters helping the ultra-nationalists in Ukraine, according to Mamon, was to use Ukraine as a base for reigniting a new front in the Caucasus. This could eventually become a security threat for Europe and Asia. The USA and Britain were among nations that have previously been implicated in the movement of Jihadists to the post-Soviet world (Ahmed 2006). The Russian President Putin claimed that Russian security services had recorded direct contact between North Caucasus fighters and representatives of US intelligence in Azerbaijan, as he revealed in the documentary, *Crimea: The Road to the Motherland*, released by the Rossiya-1 channel (Nazemroaya 2015). Volunteer battalions with significant numbers of Jihadists included: (1) the Dzhokhar Dudayev Battalion in the conflict zone in the Donbass; (2) the Sheikh Mansour Battalion, based around Mariupol and splintered off from the Dudayev Battalion; (3) the Crimea Battalion, based in Kramatorsk; and (4) a company of Crimean Tatar fighters. In the wake of worsening relations between Turkey and Russia, over Russia's continuing support for the Assad regime in Syria, Thierry Meyssan claimed the existence of a deal between Ankara and Ukraine to expedite the travel of Islamist fighters for the purpose of destabilizing Russia's hold of Crimea (Meyssan 2015). A report emanating from the *Kuwait News Agency* in November 2015 suggested that Chinese-manufactured missile launchers or

MANPADS were being sold from Ukraine to ISIS in Syria, via Turkey. Analyses of this report appearing in the *Sofia Globe* and *Vice News* were highly skeptical (Vlasova 2015; Losh 2015b).

A purportedly internal document produced by Ukraine's Southern Region Military Prosecutor, Pavel Bogutsky, and leaked in July through CyberBerkut, a pro-Russian hacking group, indicated that such militia had turned into organized criminal gangs, acting with grand impunity. A registry of crimes subject to pretrial investigation included 200 "illegal actions by territorial defense battalions and other military units involved in the Anti-Terrorist Operation in the Luhansk Region," including armed assault, robbery, kidnapping and murder. Battalions did not follow orders from the regular Ukraine army with whom they were supposedly fighting. The document suggested the Security Service of Ukraine (SBU) was also implicated (Peters 2015b).

In early July 2015, the Right Sektor, now affiliated with Islamic Jihadists, demonstrated in Kiev under the banner of "The March of the Volunteers in Honor of Prince Svyatoslav the Brave," a reference to the tenth-century military conqueror who had sought to expand Kiev's control over the East and was now a hero of Ukrainian nationalists. A solidarity march was held in Zaporizhia, a southeastern city along the banks of the Dnieper and the sixth largest in the country.

The Right Sektor demanded that the Ukrainian President Petro Poroshenko end the ceasefire with ethnic Russian separatists in the country's southeast, while also insisting that Kiev provide all volunteer battalions with "weapons, appropriate rights, and social welfare benefits." The Right Sektor threatened that the government would fall very quickly if it did not accede to its demands. A shoot-out just a few days later between Right Sektor militants and police in the town of Mukachevo had left several police and civilians dead and may have been related to a turf war over smuggling routes (Reuters 2015e). The Right Sektor spokesmen called for the sacking of the Interior Minister Arsen Avakov and demonstrated noisily in Kiev. The President Petro Poroshenko ordered the nation's security services and police to organize a disarmament of various far-right "illegal groups," saying they would not be allowed to destabilize the country. With reference to the Right Sektor, he said he did not believe any political faction in Ukraine should have its own armed "criminal cells" (Ditz 2015c). Dmytro Yarosh, the head of the Right Sektor, demanded amnesty for his fighters in Mukachevo, who were still blocked in the woods in Zakarpattia, in exchange for his organization joining the ranks of the Ukrainian Army (on their terms) (UNIAN 2015d). At an ensuing Right Sektor rally in Kiev on July 21, Dmytro Yarosh called for a referendum to impeach President Petro Poroshenko and his government. He called for the recognition of volunteer battalions and their right to carry arms as well as the introduction of martial law. The outcome of the Mukachevo clash did not bode well for the Poroshenko government:

> The aftermath of the Mukachevo standoff was a clear victory for the rightists, who saw their leader, Dmytro Yarosh, a member of parliament,

> negotiating with the Interior Ministry – and Right Sector militia blocking the road from Kiev to the scene of the fighting. The result was an announcement from the Interior Ministry that the police chief of Mukachevo has been suspended, pending an "investigation" of the charges of aiding and abetting smuggling.
>
> (Raimondo 2015b)

This was at least the second time Poroshenko had found himself in direct confrontation with fascist militia. Only months beforehand militias had felt encouraged by Avakov (Carden 2015) – whose resignation they would be demanding in July – since Avakov had a long-standing alliance with Azov and was a key supporter of Prime Minister Arseniy Yatsenyuk. Azov's leader, Andriy Biletsky was a Member of Parliament. When President Poroshenko was rumored to be considering the disbanding of the Aidar Battalion in early February 2015, militias marched on and surrounded the Ministry of Defense. In September 2014, Amnesty International issued a report on the crimes of the Aidar Volunteer Battalion. Later in the year, Aidar was a major factor in attacks on residents and police of the Kalanchak, Chaplinsky and Genichesky districts bordering Russian Crimea, requiring the formation of local self-defense units in a climate of increasing lawlessness (*Sputnik News* 2015b). In December 2015, Ulrich Heyden reported in the German newspaper *Telepolis* that in Ukraine "there is almost daily right-wing violence against the police and judiciary" and he cited many attacks by the Right Sektor against judges and high government officials throughout the country. For example, on November 30, 100 members of the Right Sektor stormed the Malinowski Court in Odessa:

> Security officers were simply pushed aside. Masked men and muscular women stood threateningly before the three judges. The judges had approved a ruling to release on bail five people detained since the violent events of May 2, 2014 in the city. The judges were threatened by masked vigilantes if they didn't sign letters of resignation. A video report captures the confrontation. It shows the anxious looks on the judges' faces. They signed the resignation letters and exited the courtroom.
>
> (Heyden 2015b)

As many as thirty such private volunteer militias were reported by Taub (2015) – among them Azov, Right Sektor, Aidar, Donbass, Dnepr-1 and Dnepr-2 – most of them benefiting from the patronage of oligarchs such as Ihor Kolomoiski (governor of Dnipropetrovsk in Eastern Ukraine to March 2015) and drawing from a long Ukrainian tradition of political parties allying with armed paramilitaries. While most groups nominally reported to the Ministry of the Interior or the Ministry of Defense, they could not be relied on to follow orders from the regular Ukrainian army, and might at any time succumb to the ambitions of personal power. Hence, there was pressure to absorb paramilitaries within the main army or in other ways to integrate these two structures of military power.

Dmytro Yarosh, the founder and leader of the fascist Right Sektor organization, was appointed official advisor to the commander-in-chief of the Ukrainian army in April 2015. In addition, the paramilitary militias of the Right Sektor were integrated into the armed forces and Azov was given permission to increase from 400 to 2,000 fighters. This does not seem to have precluded the potential for clashes between the militia and regular military or law enforcement agents, as the events of mid-July suggest.

On June 10, even the US House of Representatives voted unanimously to bar US assistance to the Azov Battalion on account of its Nazi links. But under later pressure from the Pentagon, Congress stripped the 2016 spending bill of an amendment that would have prevented funds from falling into the hands of Ukrainian neo-fascist groups. The bill included "$64 billion for overseas contingency operations" for, among other things, assisting "European countries facing Russian aggression." The Pentagon claimed that the neo-fascist amendment was redundant because similar legislation – known as the Leahy Law – already existed that would prevent the funding of Azov. On further investigation, Carden determined that the Leahy Law covered only those groups for which the "Secretary of State has credible information that such unit has committed a gross violation of human rights." The State Department had never claimed to have such information about Azov. In any case, the Leahy Law could not block funding to groups in the absence of "credible information" that they had committed human-rights violations (Carden 2016).

The struggle over the neo-fascist amendment was belated evidence that Western elites could now bring themselves to acknowledge the presence of fascism in Ukraine. However, the phenomenon of irregular militias funded by and reporting to oligarchs was not confined only to the right. President Yanukovych, who had been overthrown in the coup of February 2014, was supported in Eastern Ukraine by one such group, Oplot whose leader, Aleksandr Zakharchenko, became prime minister of the separatist Donetsk People's Republic.

Events in March 2015 confirmed the view that the *coup d'état* of February 2014 was little more than a reshuffling of power in a competition among Ukraine's leading oligarchs who had financed the uprising, and all of whom shared the same basic interest namely, to perpetuate chaos, the prime condition which had enabled them to grow their wealth (Bullough 2014). The oligarchs had made their money from fraud and coercion, seizing what had previously been State assets of the Soviet Union. On March 25, 2015, President Petro Poroshenko sacked the oligarch Kolomoiski from his position as governor of Dnepropetrovsk. A few days' previously, Kolomoiski's paramilitaries had stormed the headquarters of a Ukrainian oil company and an oil pipeline company. These were majority State-owned companies in which Kolomoiski held significant stakes. Kolomoiski was reacting against Poroshenko's attempt to wrest management control back to the control of the State. Four backers of Kolomoiski in the Ukrainian Parliament resigned from Poroshenko's Bloc, while two members of Poroshenko's cabinet were arrested for corruption. The prime minister, Arseniy Yatsenyuk, was also under parliamentary investigation for corruption. Yatsenyuk had been sponsored by two oligarchs – Dmytro Firtash and Victor Pinchuk – who in that same week attended a conference in Vienna titled "Ukraine Tomorrow," that in turn was sponsored by several EU politicians. Firtash, who had earned Nuland's ire for supporting Vitaly Klitschko to replace Yanukovych over Nuland's preference for Yatsenyuk, was facing bribery charges in the USA (Van Auken 2015b). Attempts by the USA to extradite Firtash from Austria were eventually dashed in April 2015 by an Austrian court, whose judge expressed extreme skepticism over the merits of the case (Cockburn 2015).

The evidence and threat of fascism in the Ukrainian *coup d'état* of February 2014 is hardly trivial. If it was underestimated by some commentators, this was not because it was not important or not serious. Insufficient weight was given to the strength of popular discontent with political corruption among middle-class Western Ukrainians for whom an agreement with the EU may have seemed a liberation, but whose discontent was fodder for far more sinister forces. Maidan achieved political change in Kiev and throughout Ukraine because of organized right-wing militant groups. Fascist forces were strong in the new Ukraine and although they did not dominate the political apparatus (the impact of Svoboda diminished significantly 2012–2015 from thirty-six seats in the Rada to only six), their behavior was a constant and intimidating threat to the State which in many ways depended on them for Kiev's war with the Donbass, and for the securing within Kiev's grasp of Odessa and Mariupol. An organized fascist sector that for well over a year scarcely earned acknowledgement or regret from Washington accentuated the depth of provocation that Maidan represented for Russia, and helped reignite a climate of hostility between Russia and the West that has enhanced the threat of nuclear annihilation for untold millions of the earth's inhabitants.

# 4 Absent context (1) struggle for Eurasia

Absent from WMM coverage overall was a mature appreciation that the Ukraine conflict implicated hostility between the two largest nuclear powers. The MIT missile expert, Theodore Postol, detected a sophisticated effort to ready US nuclear forces for direct confrontation with Russia. Postol (cited by Deutsch 2015) judged that technical disparities increased the danger of nuclear war. The Russians lacked a working, space-based, early warning system that could detect incoming missiles below the earth's horizon. They had six minutes to decide whether to counter-attack with nuclear weapons. The veteran foreign policy consultant and Kennedy adviser during the Cuban missile crisis, William Polk, worried about the escalation of nuclear tensions and the multiple possibilities of technical error that could ignite all-out war amidst likely fallibility of decision-making, especially in times of crisis (Polk 2015).

In 2015, the two countries each engaged in military exercises of a scale that could precipitate actual conflict. Russia launched a "snap exercise" in March involving 80,000 troops, extending from northern Russia heading southward. NATO's "Allied Shield" exercises in June involved 15,000 troops on naval vessels in the Baltic Sea, on the ground in Lithuania, Latvia, Estonia and Poland, and in warplanes over the whole region. US-NATO forces had conducted 162 exercises in 2014, double the number originally planned. The following February, NATO member states in Europe conducted an additional forty exercises in Narva, a town on the Estonian–Russian border, involving "the presence of armored vehicles from the US Army's Second Cavalry Regiment, as well as British, Dutch, Spanish, Lithuanian and Latvian troops at a military parade commemorating Estonia's Independence Day" (Martin 2015, citing a report of the European Leadership Network). In September, NATO opened "NATO force integration units," each manned by some forty staff in Bulgaria, Estonia, Latvia, Poland and Romania, in the alliance's largest reinforcement of collective defense since the end of the Cold War (Dapkus and Keyton 2015). Moscow expressed concern that month at US plans to modernize twenty nuclear weapons stationed at a German airbase – representing the newest upgrade to the B61 air-dropped nuclear bomb. The USA was modernizing the B61 arsenal with a $1.2 billion allocation from the US defense budget, part of a nuclear modernization policy first outlined in the 2010 US Nuclear Posture Review. The *Moscow Times* reporter, Matthew Bodner, cited the Russian spokesman, Dmitry Peskov, as calling the move a potential "violation of the strategic balance in Europe," that would demand a Russian response. Expert sources considered the response might be the redeployment of Iskander-M tactical missiles to Kaliningrad, a Russian enclave in Eastern Europe bordering Poland and Lithuania. Neither the Iskander-M nor the US B61 bombs were governed by arms control treaties. The USA continued to deploy tactical nuclear weapons in NATO countries such as Germany, Belgium, the Netherlands, Italy and Turkey. Russia was pursuing a fully-fledged overhaul of its nuclear forces (Bodner 2015c). NATO's opening of new offices in Eastern European countries added to the perception of a NATO intent on encircling Russia. Ukraine fed this perception by participating in NATO war games.

Such escalation confirmed the perils of the Ukraine dispute and called for heightened sensitivity by WMM publishers to the inadvisability of standard war reporting frames oriented towards displays of war and violence, elite interests, and the prospects for victory (zero-sum), as against alternative approaches to conflict coverage that emphasize multiple perspectives, the interests of non-elites, and efforts towards peaceful conflict resolution (Lynch and McGoldrick 2005). Boardman summarized the Western contribution to an escalation of tensions with Russia as follows:

> Poland, Norway, and the Netherlands (all NATO members) are expanding their submarine fleets, continuing the militarization reaction to Crimea. The US Navy has increased its presence in the Black Sea as an anti-Russian move. On September 1, the US Navy started operation Sea Breeze, joint operations with the Ukraine Navy (which lost more than half its ships in Crimea) and nine other countries [...] Ukraine has used internationally-banned cluster bombs against the separatists in the Donbass, one of five countries in the world where they have been used this year (the others are Yemen, Syria, Sudan, and Libya). The US, which has not signed the treaty banning cluster bombs, continues to make and distribute these anti-personnel weapons that take a heavy toll on civilians [...] Ukraine has renewed its bid to join NATO.
>
> (Boardman 2015b)

Russian scale as a military power contributed to the warrior passions of some US neo-conservatives whose interests generally aligned with those of the US military-industrial complex (Savage 2015). A big opponent (but not too big) provides a strong pretext for bigger budgets, quite attractive therefore when compared to smaller fry preoccupying the USA in the Middle East and South West Asia.

## Long heritage

WMM downplayed centuries-old bonds between Russia and Ukraine, and the equally established tension internal to Ukraine between an orientation towards Europe and a contrasting orientation towards Russia. Ties between Ukraine and Russia are extensive. Successors to Ryurik, a legendary Varangian prince invited by the Northern tribes to rule over them, established Kiev as their capital city. Thus Kiev was capital of the first "Russian" state in the ninth century. Pre-modern Russia was sometimes described as "Kievan Rus."

One-third of Ukraine's people were Russian-speaking and bilingual. Demands for a democratic federation, reflective of Ukraine's ethnic diversity yet independent of Moscow, had long pedigree in Ukraine. Pilger notes of Eastern Ukrainians that:

> Most are neither "separatists" nor "rebels" but citizens who want to live securely in their homeland. Separatism is a reaction to the Kiev junta's attacks on them, causing as many as 110,000 (UN estimate) to flee across the border into Russia. Typically, they are traumatized women and children.
>
> (Pilger 2014)

## Western destabilization

Most day-by-day WMM coverage reported the conflict without reference to evidence of medium- to long-term Western strategies for destabilization of Russia's relationship to Ukraine specifically, for establishing Western hegemony over Eurasia more broadly. Chris Marsden noted:

> The US has spent the past two decades seeking to eliminate Ukraine as a strategic buffer between Russia and the West, sponsoring the "Orange Revolution" in 2004 in an ultimately abortive attempt to install a wholly pro-Western government. Washington and its allies have tried to do the same in other former Soviet states by integrating them into the structures of NATO and the European Union, encouraging Georgia, in particular, and former Soviet republics in Central Asia to take the path of Estonia, Latvia and Lithuania.
>
> Washington has been funneling money into the region for years and has now opened the taps all the way. According to an admission in December by Assistant Secretary of State for European and Eurasian Affairs Victoria Nuland, the US had invested "over $5 billion" to "ensure a secure and prosperous and democratic Ukraine."
>
> (Marsden, cited by Cromwell and Edwards 2014)

The prominent neo-conservative Victoria Nuland – once national security adviser to the former US Vice-President Dick Cheney – was the wife of the co-founder of Project for the New American Century (PNAC), Robert Kagan. In a leaked conversation between Nuland and Geoffrey Pyatt, US Ambassador to Ukraine in February 2014, and later authenticated by Washington, Nuland identified Yatsenyuk as the US-preferred candidate for prime minister, while profanely dismissing the interests of the EU. Sakwa read in this a "high degree of US meddling in Ukrainian affairs, and the way that the concerns of its ostensible allies and partners are dismissed with a profanity" (Sakwa 2015: 87).

WMM, attending principally to elite Western sources, were unreasonably insensitive to – if not scathing of – Russian perspectives. Within 100 years Western powers had invaded Russia three times – the two world wars and the "Intervention" of 1918–1920 – using Eastern Europe as a highway on each occasion (Blum 2014). Was it any cause for wonder, Blum asked, that after World War II the Soviets wanted to shut this highway down? Following World War I, Western powers had helped detach the Baltic countries – components of the Russian empire from 1721 – from the Soviet regime. The West allowed defeated Germans to continue occupying the Baltics, allowing time for the installation of collaborator governments that soon declared independence from Russia.

The Liberal analyst, Stephen Walt (Walt 2015), in seeking to break through what Robert Parry (Parry 2014r) described as the groupthink of the Washington policy circle, argued that the USA was applying the wrong model. Washington advocates for US arming of Kiev were mistakenly applying a "deterrence model" to what was in fact, Walt explained, a "spiral model" situation. Russia was not pursuing imperialist expansion but was rather a weakened ex-superpower fearful of further encroachment against its sphere of influence. Walt's perspective placed Washington in the role of an incompetent but fundamentally well-intentioned power and as such blacked out a broad canvas of important, deep-seated, and long-term dimensions to the conflict to which I shall now turn.

## Greater Europe and Silk Roads

A broader panorama of tension between Russia and the West involved two "greater-Europe" projects (Sakwa 2014). The first, promoted by the Brussels-centered EC, was cemented through a range of initiatives including the "Eastern Partnership" formed in 2008. This co-opted six of Russia's ex-Soviet republics into the Western sphere. An alternative vision of a Eurasia stretching from the Atlantic to the Pacific and centered in Moscow was preferred by Russia.

Washington's enthusiastic support for the coup regime of Ukraine and the consequent agitation of Russia reflected US fears that Eurasia had shifted away from the West in favor of positive relations between Russia, China and, through the China-based Asian Infrastructure Investment Bank (AIIB and the Shanghai Cooperation Organization), Iran, India, Pakistan and many countries inside or in the immediate proximity of Eurasia. US policy sought to demarcate dollar zone economies from the BRICS and in particular the Russian and Chinese economies. This threatened to undermine Russian and Chinese power as creditors to Third World countries by excluding them from favorable access to IMF debt judgments in the event of non-payment of loans:

> Moving to denominate their trade and investment in their own currencies instead of dollars, China and Russia are creating a geopolitical system free from US control. After US officials threatened to derange Russia's banking linkages by cutting it off from the SWIFT interbank clearing system, China accelerated its creation of the alternative China International Payments System (CIPS), with its own credit card system to protect Eurasian economies from the shrill threats made by US unilateralists.
>
> Russia and China are simply doing what the United States has long done: using trade and credit linkages to cement their geopolitical diplomacy. This tectonic geopolitical shift is a Copernican threat to New Cold War ideology: Instead of the world economy revolving around the United States (the Ptolemaic idea of America as "the indispensable nation"), it may revolve around Eurasia. As long as the global financial papacy remains grounded in Washington at the offices of the IMF and World Bank, such a shift in the center of gravity will be fought with all the power of the American Century (indeed, American Millennium) inquisition.
>
> (Hudson 2015b)

Washington's anxiety helped explain a sharp division of attitudes in Washington towards Iran. One faction was inclined to the view that Iran could replace Russia as a source of oil and gas for Europe, thus heading off a unified Euro-Russian power bloc. A rival faction continued to regard Iran as a threat to US and Israeli dominance of the Middle East, one that must be suppressed at all costs (Whitney 2015b). The first tendency helped explain why the USA applied pressure – fruitlessly – on both Germany and France to discourage China from embracing Russia as well as Germany in its vision for the "New Silk Road" and "Maritime Silk Road":

> The land-based "New Silk Road" will begin in Xi'an in central China before stretching west through Lanzhou (Gansu province), Urumqi (Xinjiang), and Khorgas (Xinjiang), which is near the border with Kazakhstan. The Silk Road then runs southwest from Central Asia to northern Iran before swinging west through Iraq, Syria, and Turkey. From Istanbul, the Silk Road crosses the Bosporus Strait and heads northwest through Europe, including Bulgaria, Romania, the Czech Republic, and Germany. Reaching Duisburg in Germany, it swings north to Rotterdam in the Netherlands. From Rotterdam, the path runs south to Venice, Italy – where it meets up with the equally ambitious Maritime Silk Road.
>
> The Maritime Silk Road will begin in Quanzhou in Fujian province, and also hit Guangzhou (Guangdong province), Beihai (Guangxi), and Haikou (Hainan) before heading south to the Malacca Strait. From Kuala Lumpur, the Maritime Silk Road heads to Kolkata, India then crosses the rest of the Indian Ocean to Nairobi, Kenya. [...] From Nairobi, the Maritime Silk Road goes north around the Horn of Africa and moves through the Red Sea

into the Mediterranean, with a stop in Athens before meeting the land-based Silk Road in Venice.

(Tiezzio 2014)

Symptomatic of the challenge to the USA to fend off a Russian–European rapprochement was the decision in April 2015 of the French oil giant Total to progress a $27 billion project to develop a liquid natural gas (LNG) terminal at Yamal, near the Arctic Ocean in Siberia. Because sanctions against Russia limited Total's capacity to finance the project from US or European banks, the corporation decided to look to Chinese banks for up to $15 billion.

Washington anxiety was amplified, Escobar noted, as the realization sank in that: (1) sanctions had not crippled and in some respects may have strengthened Russia; (2) Putin's statesmanship was unequalled by his counterparts in the Western world, demonstrating greater success than any in bridling the ambitions of corporate or oligarchic power; (3) the peoples of Crimea, Donetsk and Luhansk had successfully repudiated Kiev fascism; and (4) Russia's 10,000 tactical nuclear weapons, S-500s, anti-missile missiles and anti-aircraft missiles could take down anything the West might fire on Russia and could penetrate any Western defense (Escobar 2015b).

The achievement of a Eurasia compatible with US hegemony would exploit the political weakness of corrupt oligarch regimes to extend NATO and incite actual, crippling conflict between Russia and Europe without Chinese intervention and from which the USA would emerge relatively stronger.

## US betrayal and the Eastern buffer

Equally absent from WMM accounts were American promises to former Russian president Mikhail Gorbachev that after the dissolution of the Soviet Union, NATO (70+ percent financed by the USA) would not expand its borders to the east. The former CIA analyst Ray McGovern (2014a) cited Jack Matlock, the then-US ambassador to the USSR, confirming that the basic original agreement involved: (1) Gorbachev's pledge not to use force in Eastern Europe where the Russians stationed twenty-four divisions (some 350,000 troops) in East Germany alone; and (2) Bush's promise not to "take advantage" of a Soviet withdrawal from Eastern Europe. Matlock, whom McGovern regarded as a widely respected Russia expert, told him that "the language used was absolute, and the entire negotiation was in the framework of a general agreement that there would be no use of force by the Soviets and no 'taking advantage' by the US" (McGovern 2014a). Nevertheless, under the Clinton and Bush II administrations, "former Soviet bloc countries such as Poland, Czechoslovakia, Romania and the Baltic states were asked to join the western alliance, and they did" (Johnson 2014). By the 2000s NATO's borders were much farther east and Russia was concerned for the security of its fleet in the Crimean (Johnson 2014). Bill Clinton initiated this process, not to be outdone by his rival for the presidency, Bob Dole, in the 1996 elections. In 1994, Clinton publicly supported the idea of expanding NATO membership. Applicants had to be democracies willing to put their forces under NATO command. The EU had similarly seen significant expansion, with ten new members joining in 2010, and an "Eastern Partnership" framework provided for non-eligible post-Soviet states.

A goal of US support for the 2004 color revolution was Ukraine's incorporation into NATO. Much progress had been accomplished in this direction before 2004, even more before 2014. Anti-Russian Western aggression from 2014 was less a reaction to a perceived and unanticipated frustration of NATO's legitimate interest but more likely indicated a broader strategy of conquest. The scholar Stephen Cohen wondered in an interview with Patrick Smith (Smith 2015a) whether the Cold War had ever actually ended and whether Russia could ever be entitled to a sphere of interest in a world where the USA considered that only the USA could have one.

Summiting in Bucharest in April 1998, NATO leaders declared that "Georgia and Ukraine will be in NATO." Washington then launched the United States–Ukraine Charter on Strategic Partnership. Under President Yushchenko, Ukraine sought to temper Russian influence in Crimea, reserving the right to prevent Russian vessels from using the Black Sea Fleet base in Sebastopol. Ukraine participated in the US-led NATO Partnership for Peace military exercises every year from 1996, except 2006, and collaborated with the Obama administration's European Phased Adaptive Approach missile shield being deployed along Russia's western and southern borders (Rozoff 2014). Rapid Trident military exercises took place annually in Ukraine in collaboration with US, NATO and Partnership for Peace forces. In 2010, Ukraine contributed a warship for NATO's Operation Active Endeavour in the Mediterranean and in 2012 provided a warship to NATO's Operation Ocean Shield off the Horn of Africa.

Russia's foreign ministry warned in January 2008 that NATO expansion "could produce a serious political-military upheaval that would affect the interests of Russia" (cited in Sakwa 2015: 54). The 2008 Russo-Georgian War became the first of the wars to block NATO enlargement. Victoria Nuland unsuccessfully urged President George Bush to grant NATO membership to both Georgia and Ukraine. For Yushchenko, NATO membership was considered the only guarantee of Ukrainian sovereignty and territorial integrity (Sakwa 2015: 54–55). Yushchenko exacerbated tension with Russia when he reaffirmed that basing rights for the Russia's Black Sea Fleet at Sebastopol would be terminated in 2017. For Yuschenko, any rapprochement between Washington and Moscow was a threat (Sakwa 2015: 70). Nonetheless, in 2010 Russia's President Medvedev and Ukraine's President Yanukovych extended the lease, granting Russia the use of Sebastopol until 2042, with the option of a further five years.

The components of the USA's betrayal of Russia were identified by Roberts (2014a): The USA (1) violated its agreement not to take NATO into Eastern Europe; (2) withdrew from the ABM treaty which had underpinned nuclear security between the great powers; (3) placed US anti-ballistic missile bases on Russia's border; (4) changed its war doctrine to permit pre-emptive nuclear attack; (5) financed "color revolutions" in Georgia and Ukraine; (6) ignored

Russian interests in the Balkans and dismembered Serbia; (7) trained and equipped Georgia's military and encouraged Georgia to attack the Russian peacekeeping force in South Ossetia; (8) instigated the overthrow of the democratically elected government of Ukraine and installed a puppet government. Other developments included the instigation of civil war in Russian ally's Syria, where it supported Jihadist rebels of whom some were linked to Al Qaeda and, later, to ISIS, in attempt to topple the Syrian leader, Assad.

## Corruption

The discourse-of-pretext deployed by EuroMaidan protestors and their Western supporters for the violent ejection of President Yanukovych in 2014 alleged that he was "corrupt." WMM made much of the apparent wealth on display in the presidential residence. This judgment ignored the context of recent Ukrainian history and contrasted lamely with the excesses of billionaires almost anywhere and those of the candidate who was later elected to replace Yanukovych, chocolate candy manufacturer Petro Poroshenko.

> Without a person's knowing that Ukraine is by far the most corrupt country in the former Soviet Union, and the one with the worst economic performance of them all, Ukraine's politics just can't be understood at all: it has long been an extreme kleptocracy, ruled by psychopathic politicians, for the benefit of psychopathic oligarchs, who have robbed the country blind.
>
> (Zuesse 2014a)

The impeachment procedure against Yanukovych did not comply in any significant way with the Ukrainian constitution. Even the vote of MPs that did take place failed to reach the required majority. Corruption within Ukraine's senior establishment was not an obstacle to US and NATO extension of influence and would accommodate to what Rasmus described as a twenty-first-century phenomenon of "colonialism by means of financial asset transfer" – a process of wealth extraction by "colonizing country managers, assigned to directly administer the processes in the colony by which financial assets are to be transferred" (Rasmus 2015). This took three main forms: interest payments on ever-rising debt; fire sales of government assets sold directly to the colonizer's investors and bankers; and co-option of the colony's banking system and bank assets for transfer of wealth to shareholders of the colonizing country's private bankers and investors.

There was a great deal of anti-corruption rhetoric both in the new regime and among its Western supporters but it had lost credibility with the Ukraine public by the end of 2015 when Poroshenko rated less popular than Yanukovych at his lowest. It was undermined by the continuing presence of corrupt politicians in the Rada and by problematic Western links to Ukrainian corporations – including the involvement of the US Secretary of State Joe Biden's son on the Board of a company owned by one of Ukraine's biggest oligarchs.

## Chaos theory

Among the few commentators to consider the "long game," Pepe Escobar (author of *Globalistan*, 2007) turned his spotlight on the Russiaphobe policy of the US ex-foreign policy advisor, Zbigniew Brzezinski. Brzezinski had long advised against allowing any Russia–China alliance; putting a stop to any German–Russian alliance; and securing Eurasia under Western hegemony. This confirmed the assessment of Immanuel Wallerstein that, for the "Empire of Chaos," restoring its hegemony in the world-system had become the supreme obsession. An immediate goal, for Brzezinski, was Russia's energy wealth:

> The Dr. Zbig "Grand Chessboard" Brzezinski-style strategy has always been to lure Russia into another Afghanistan in Ukraine, leading to a collapse of the Russian economy with the Big Prize being a Western takeover of Russia's oil and natural gas wealth, and by extension Central Asia's. Ukrainians would be used as cannon fodder, as were Afghans since the 1980s Arab-Afghan jihad.
>
> (Escobar 2015a)

Equally in tune with the premise of "chaos theory," was the postulation that:

> The deeper the US slips into economic decline, the higher it ratchets up the pace and stakes of armed conflict. Washington appears to have crossed some kind of Rubicon, "to embark on a mad, scorched earth policy to terrorize the planet into submission through relentless escalation into a global state of war."
>
> (Ford 2014)

Putin's advisor and friend Sergei Glaziev (Whitney 2014f) argued that the huge outflow of capital from Europe following World Wars I and II had transformed the USA into a superpower and now, faced with economic decline, the USA was trying to precipitate another European war in order to achieve the same objective of an intercontinental flow of capital in the US favor. Threatened by the transition from an era of American capital accumulation to an era of Asian capital accumulation, the USA regarded war in Europe as the best way to preserve its hegemony. The USA was threatened by increasing dependence of Europe on sources of oil and gas originating from or controlled by Russia and Iran, soon to be delivered by a pipeline (the PARS pipeline) under construction from Iran, running through Iraq and Syria, and outside of the prevailing petrodollar system. This system had required OPEC members to trade in dollars and recycle their profits in US Treasuries and other dollar denominated assets, which in turn had propped up the USA's debtor status. This would be undermined by the rise of trade in alternative currencies.

Consistent with Glaziev's assessment was that of Mahdi Darius Nazemroaya who considered that the ultimate goal of the USA was to prevent any alternative powers from emerging in Europe and Eurasia that could threaten Euro-Atlantic integration, and to achieve this by the same policy of fractionalization, as mooted by Zbigniew Brzezinski (once US President Jimmy Carter's national security advisor and architect, through provocation, of the Soviet invasion of Afghanistan). Brzezinski's vision was of a loosely confederated Russia, composed of a European Russia, a Siberian Republic and a Far Eastern Republic (Nazemroaya 2014b). These components would then be more amenable to closer economic relations with Europe, Central and East Asia. Nazemroaya cited Dmytro Sinchenko, a Ukrainian analyst and Radio Liberty contributor, who advocated reinvigoration of the concept of a pro-US Georgia-Ukraine-Azerbaijan-Moldova (GUAM) Organization for Democracy and Economic Development that would join NATO and assault Russia. The end goal:

> is to divide (balkanize) and pacify (finlandize) the world's biggest country, the Russian Federation, and to even establish a blanket of perpetual disorder (somalization) over its vast territory or, at a minimum, over a portion of Russia and the post-Soviet space, similarly to what is being done to the Middle East and North Africa. The future Russia or the many future Russias, a plurality of weakened and divided states, that Washington and its NATO allies see is/are demographically in decline, de-industrialized, poor, without any defensive capabilities, and hinterlands that will be exploited for their resources.
>
> (Nazemroaya 2014b)

## Energy and energy prices

In 2014, the EU imported 30 percent of its natural gas requirements from Russia, and about half of that came via Ukraine. Up to 2014 over 75 percent of Russia–China trade settlements were paid in dollars. Moscow persuaded China and Iran to drop the dollar in their bilateral dealings. Other trading partners were eager to do the same. Fear of de-dollarization had contributed in part to US support for the destabilization of Syria, where Assad had rejected a Qatari proposal for an alternative pipeline route. Completion of Russia's preferred PARS pipeline would reduce US regional influence and hurt the interests of US allies such as Israel, Saudi Arabia, Qatar and Kuwait.

Adapting the British imperial policy of divide and rule, the USA sought to weaken Europe, which could lose one trillion dollars as a result of an anti-Russian sanctions policy imposed on Europe by the USA, while initiating the fragmentation of Russia by pitting Ukraine against Russia. This would afford the USA an opportunity to write-off its public debt to these countries. The implications partially surfaced as European support for the USA fractured in February 2015. As the USA ratcheted up pressure against Russia by threatening to arm Ukraine, European leaders were obliged to consider their costs – more severe than any that would be shouldered by the USA. Europe was simultaneously hit by the arrival of a million Middle Eastern refugees seeking shelter from devastation for which the USA and Europe bore considerable responsibility. EU–Russian trade was ten times the volume of US–Russian trade. Germany was the biggest EU loser: German exports to Russia fell by 18 percent in 2014, representing a loss of €8 billion to the German economy and costing up to 60,000 jobs (Durden 2015b; DW 2015b). Dutch exports to Russia fell by nearly 40 percent (Reuters 2015g). The Russian economy was hit not just by sanctions but also by the falling price of oil and devaluation of the ruble. A weaker Russian economy was not good news for the European.

Yet lethally destabilized Russia would ultimately benefit US and EU corporate interests and promote US hegemony in Eurasia. The US ally, Saudi Arabia, deliberately persisted with high oil production in the face of falling prices – the result of an oil glut enabled in part by alternative fuels and shale oil production, soon to be exacerbated by the return to the market of Iran, following an international deal that lifted sanctions imposed over the (manufactured) crisis of Iran's potential nuclear arms capability (Porter 2014). Lower oil prices hurt Russia, and Russia was successfully enabling Saudi Arabia's enemy, President Assad of Syria, to withstand a Western- and Saudi-instigated civil war. Iran, regional rival to the Saudis and a Russian ally, would likewise suffer. Low prices could puncture the boom in non-traditional sources of fossil fuel production that had been spearheaded by the USA. But Saudi Arabia would have to reckon with the consequences of declining revenue for the welfare and loyalty of its citizens (Klare 2016).

Engdahl claimed an explicit deal had been struck between the Saudis and the US Secretary of State John Kerry on the occasion of Kerry's visit to the oil kingdom on September 11, 2014, negatively impacting the economies of three US opponents: Iran, Russia and Venezuela (Engdahl 2015). There were previous examples of Saudi price manipulations directed against Russia: for example, in protest against Russia's invasion of Afghanistan in 1979 and against Russia's competitive threat in 1998 when Saudi Arabia halved the price of oil and forced Russia to default on its debt. As OPEC's de facto leader, Saudi Arabia mysteriously abandoned the cartel's "traditional strategy" when it announced in mid-2014 that it would not cut production, even if prices dropped to $20 per barrel (Whitney 2014h) and even though this was in a period when, given the conflicts raging in Iraq and Syria, the prices would normally be rising (Engdahl 2015). Saudi Arabia was the main player within OPEC and it was stalling attempts to cut production. The consequences for Russia of both the oil price reductions and Western sanctions – which cut Russia off from international lenders – were severe: the ruble lost over half of its 2013 value against the dollar; inflation was at 16 percent; government reserves dropped by a third in 2014–2015; the economy as a whole retracted by 4–5 percent; state employees experienced salary freezes and cuts; there was a decline in real incomes of 5 percent; purchasing power fell by over 8 percent in the second quarter of 2015 alone, year on year, with declines in retail sales and increases in prices (Oakford 2015). Hopes that oil deals with China would save Russia proved illusory as the Chinese economy also contracted.

An additional underlying tension between Russia and the USA, generally disregarded by WMM, concerned their rival ambitions in the Arctic where climate change and the thaw facilitated access to raw materials. The USA regarded the Arctic as a region in which it had fundamental national security interests. These necessitated expanded penetration by the US Navy's submarine-launched ballistic missile (SLBM) third of the American nuclear triad (US Department of the Navy, cited by Deutsch 2015). From December 2013, Russia had already embarked on pumping Arctic crude oil from its Prirazlomnaya platform, extracting more than 4 million gallons by summer 2015. Further Russian progress was hampered by the US-imposed sanctions that extended to the export of high technology oil equipment to Russia (*Moscow Times* 2015a).

## International law and the media

Many measures taken by the USA, NATO and EU vis-à-vis Ukraine were contrary to international law. External interference in the internal affairs of a sovereign state is generally proscribed by many instruments of international law, not least among them being the 1648 Treaty of Westphalia and UN Charter. Three UN General Assembly resolutions prohibit non-consensual belligerent intervention. But the only international laws that seemed to excite WMM were those they could (mis)interpret in the case of Russia's annexation of the Crimea.

As the concept of "framing" suggests, a key part of the construction of any text that has a highly persuasive or propaganda function is to highlight features of its argument that are favorable to the interests of the propagandist (whose identity is often hidden behind journalistic or other spokesmen of the media through which the propaganda is presented) and to background or omit features that are unfavorable. There is an almost infinite scope for framing through highly selective choices of contextual information. The provision of context is intended to enable readers or viewers to better understand ongoing events; the selective manipulation of which parts of the context to supply is intended to ensure that readers or viewers understand the ongoing events in the way that is most useful to the propagandist(s). The principal element of context that was backgrounded, omitted or misrepresented in the Ukraine crisis was the titanic struggle of interests and ideas between Western powers led by the USA and its allies, on the one hand, and Russia and China, on the other, for hegemony over Eurasia.

# 5 Absent context (2) preparing the neoliberal turn in Ukraine's economy

The independent investigative reporter Robert Parry (2014m) wryly observed that: "the Ukraine case has been a curious example of US-backed 'democracy promotion' – overthrowing a democratically elected leader so a coup regime could impose a 'very unpopular' austerity plan on an already suffering population." Parry's view perhaps underestimated popular frustration with oligarchs and corruption in Yanukovych's Ukraine (and of post-independent Ukraine generally) but highlighted US support for the coup and invited consideration of benefits that would accrue to financial elites from corralling Ukraine within the Western orbit. WMM coverage, still in naive "Arab Spring" mode was deceptive as to the broader political and economic stakes and consequent risks – at public expense – of war and financial collapse.

## Agribusiness

In 2013, Ukraine was the world's third largest exporter of corn and the fifth largest exporter of wheat, with arable land (32 million hectares), equivalent to one-third of all such land in the EU. Of the total, 1.5 million hectares had been acquired by foreign companies (notably, Monsanto, Cargill and DuPont), and these were added to after Yanukovych's departure, when acquisitions extended to infrastructure and shipping. The California-based Oakland Institute reported in July 2014 that the World Bank and IMF, under the terms of their $17 billion loan to Ukraine, would open Ukraine to genetically modified organisms, crops and seeds (GMOs) (Nelson 2014), urging Ukraine to deregulate agricultural land use, promote its sale to foreign investors, delete existing provisions regarding mandatory certification of food from the listed laws of Ukraine and minimize "unnecessary costs" for businesses arising from regulations on pesticides, additives and so forth (Oakland Institute 2014). Following the reconfiguration of government in December 2014 to accommodate foreign experts, the World Bank and IMF praised Ukraine for facilitating the foreign acquisition of foreign land, cutting regulations on food and plant production, and reducing corporate taxes and custom duties (Mousseau 2015).

The agribusiness behemoth and GMO promoter, Monsanto, had invested in genetically engineered grain projects in Ukraine, "seeing the country as Europe's

Achilles Heel when it comes to resisting GMOs" (Hudson 2014a). By November 2013, major Ukrainian agriculture associations proposed legal instruments for creating, testing, transportation and use of GMOs. The US–Ukraine Business Council's sixteen-member Executive Committee was packed with representatives from US agribusiness companies, including Monsanto, John Deere, DuPont Pioneer, Eli Lilly, and Cargill, all keen to introduce GMO products (Eade 2015). Also on the board of directors was USAID. Foreign companies circumvented Ukrainian law by leasing instead of purchasing land, and buying large shareholdings in Ukrainian agribusinesses so as to change these companies from the inside out. This gave them the capability not only to introduce their technologies but also to actively change Ukrainian legislation so as to facilitate their adoption. They enjoyed active US government intervention. By sponsoring the International Service for the Acquisition of Agri-biotech Applications (ISAAA), the USA influenced the US–Ukraine Business Council, endeavoring to introduce GMOs and promote the corporate takeover of Ukraine's agriculture.

## IMF loans, austerity and privatization

Yanukovych had rejected an EU/IMF loan package to the dismay of Ukrainians who imagined that acceptance would expedite progress towards a less corrupt, more European future. The Russian loan he preferred offered more money, subsidized gas prices and would not have prohibited Ukraine from additional agreements with the EU. Either way, the scale of Ukraine's shadow economy demonstrated that more effective regulation could have avoided the necessity for foreign financial intervention altogether. This represented 40–60 percent of Ukraine's economy, avoiding government regulation, taxation or observation. From 2008 to 2015, it averaged 44.5 percent of the total economy and from 2011 to 2015 its total value approximated $328 billion ($18 billion in the first half of 2015 alone). Ninety percent of the problem lay in "misinvoicing" – the misstating of the values of goods bought or sold, especially in the highly monopolized raw materials and resource extraction industries, and extending to fraudulent value-added tax refund schemes. Two million workers had dropped off the tax rolls in 2015, and another five million reported salaries lower than minimum wage, suggesting that they were being supplemented by cash payments (Rachkevych 2015).

Ukraine had accepted previous IMF loans: $16.6 billion in 2005; and $15.1 billion in 2010. Why had these not produced a more "European" Ukraine? The 2010 loan imposed a 50 percent increase in household gas prices and cut subsidies proportionately. Allowing for the imposed cut in gas subsidies the $15 billion was reduced to a net $7.5 billion advantage to the economy (Rasmus 2014). Ukraine requested another $20 billion in December 2013 and was offered $15 billion. The IMF signed off on the total package on August 29, 2014, for the first time issuing such a loan under conditions of civil war (Hudson 2014b) and possibly contravening its own rules. The loan contributed to military expenses for Kiev's attack on its eastern provinces which had previously accounted for 25 percent of Ukrainian exports, principally to Russia. Almost none of it would benefit these provinces – whose basic infrastructure Kiev had destroyed, and whose need was desperate.

Of an initial $3.2 billion tranche disbursed to the Ukrainian treasury at the start of May, part of a two-year stand-by loan agreed in April, $3.1 billion had disappeared offshore by the middle of August (Helmer 2014), indicating either that the loan was stolen by government insiders, or that the central bank simply turned the money over to the oligarchs, who ran the country's banks as though they would their own conglomerates. Investigations by Nashi Groshi, a Ukrainian anticorruption watchdog, found that $1.8 billion of the IMF loan was siphoned into a scheme involving oligarch Kolomoiski's PrivatBank, together with fifty-four offshore entities owned by forty-two Ukrainian firms. PrivatBank issued loans that were to be guaranteed by the goods that the loans were used to purchase. The offshore entities promptly ordered supplies that were prepaid into accounts at the Cyprus branch of PrivatBank. The suppliers neither delivered the actual supplies ordered nor returned the money (Cockburn 2015). Such shenanigans helped explain how, by the end of June 2015, the banking system's capital adequacy ratio was 13.8 percent, down from 15.9 percent at the end of June 2014.

The EU+US+IMF loans were conditional on: (1) 40 percent cuts to Ukraine's energy and gas subsidies – whose main beneficiaries had been ordinary people; (2) 50 percent cuts in public sector pension payments; (3) substantial increases in consumer prices; and (4) the privatization of public firms (Petras 2014). In the first quarter of 2015, the economy declined by 17.2 percent (leveling out to "only" 14.7 percent by the second quarter) on the previous year, with a fall of 25.1 percent in private investment. A major push for the privatization of approximately 350 firms was slated to begin at a US conference in October for US and European investors. This was to encompass the sell-off of one of the country's largest power-generating plants, Tsentrenergo, and six regional power generating and distributing companies (AFP 2015b). Actual progress was slower than anticipated as investors digested the poor state of repair and the threats to safety posed by government instability and conditions of civil war. Massive opposition to privatization by the people of Ukraine was especially acute in the south and east, the location of many of the largest state-owned industries. Of the Ukrainians surveyed, 49 percent believed that all of the state-run companies should remain as they were, while 30 percent considered that a few should be privatized, but that most should remain in state hands (Melanovski 2015). The country's coal industry in the Donbass region was particularly threatened by job losses as the government prepared to sell all the state-run mines by the end of 2016 and to terminate subsidies to the industry completely by 2020.

Ukraine's debt was denominated in dollars and euros, and to raise the foreign exchange that it needed to repay the IMF, it would need to sell its natural resources, in particular gas rights and agricultural land (Hudson 2014b). Oil was an important source of endemic corruption. An appreciable part of Ukraine's debt ($3 billion) was owed to Russia for oil and gas that Russia had provided at a steep discount ever since Ukraine became independent of Russia. Loan money from the IMF would be needed to service these and associated debts to Russia, loans whose repayment, under existing contracts, would be triggered if Ukraine's overall debt rose above 60 percent of its GDP – it reached 95 percent above GDP by 2015. The sum total of Ukraine's debt to Russia may have reached $38bn at that point (Hudson 2014b). In principle, further debt was incurred when Crimea voted to return to Russia, since Russia had made an advance payment to Ukraine for the continued lease of its port at Sebastopol and the money in question should have been returned.

US advisers hoped to free Ukraine from the burden of its obligation to repay. They pondered restricting Russia's legal access to British courts where the loans were registered, but feared a severe legal blowback on Western finance and legal institutions. Such politicization of lending terms, an example of what Hudson has called New Cold War geopolitics, shared with conventional warfare the aspirations of appropriation of land, raw materials, infrastructure and the purchase of banks (Hudson 2015a).

The Ukrainian non-payment for gas it had received from August 2013 triggered an end to the long-standing Russian discounts in March 2014. The Kiev government depicted Russia's legitimate reaction as rapacious Russian price escalation. The Western instigation of non-payment of gas arrears and cessation of military exports to Russia aimed to detach Ukraine from the Russian orbit, leading to an all-out trade war by 2016 in line with US neo-conservative aspirations to stomp on any possibility of Eurasian integration between the EU, Europe and Russia. German and EU compliance (EU capitalists stood to gain from fire sales of Ukrainian assets) grew shakier from March 2015 as the EU assessed the implications of active NATO–Russian hostility. But the reduction in the market value of Ukraine's military equipment factories – a consequence of civil war in the Donbass and loss of Ukraine's Russian market – rendered these all the more appetizing to Western speculators, while Ukrainian dependency on IMF-World Bank credit deepened. All the while, Western sources blamed Ukraine's desperate economic situation on Russia, rather than on the IMF, and ignored the damage their tactics had inflicted on Russia.

In February 2015, the IMF replaced its two-year stand-by loan of April 2014 in response to the accelerating weakness of Ukraine's economy. $4 billion was extended from the original loan package, and a further $17.5 billion committed (with as much as $10 billion expected during the first year). This in itself was insufficient. Additional commitments of $2 billion from the EU, $2 billion from the USA and the anticipated agreements of bondholders to reduce borrowing costs – though many, including Russia with its $3 billion worth of Ukrainian bonds, indicated they would resist such a "haircut" – would bring Ukraine a total of $40 billion over four years. A World Bank loan of $500m agreed in September 2015 functioned as a stop-gap measure to balance the State budget. In October, on the eve of nationwide elections, the US Secretary of Commerce, Penny Pritzker, announced that the USA would provide a third $1 billion loan as soon as November, and committed the US government and American business to partnership with Ukraine "on the road to economic reform" (Williamson

2015a). In November, the European Investment Bank (EIB) provided financial guarantees for $500 million loans from the World Bank to the Ukrainian gas company, Naftogaz, with the funds to be used for gas purchases from Russia's Gazprom for the 2015–2016 winter season.

The February 2015 IMF agreement followed on a State austerity package passed in December 2014. This reduced the population of public sector workers by 10 percent, partially privatized health (cutting public expenditure by 40 percent) and education (cutting public expenditure by 20 percent and eliminating two years of free public schooling), removed price controls on medicine, sold off or privatized the nation's coal mines, and eliminated the remaining price controls and subsidies on consumer gas – leading to price increases of three to five times the going rate (Williamson 2015a; Lantier 2015c). The average monthly salary in Ukraine in early 2015 stood at $170 and real disposable income had fallen 24 percent year-on-year by mid-2015 (see Klikushin 2015 for discussion of trends). The official rate of inflation was 30 percent (or 272 percent after currency devaluation was taken into account), climbing to 46 percent by mid-2015 (US–Ukraine Business Council 2015).

The Ukrainian government and the IMF agreed that the current debt load of 95 percent of GDP was unsustainable and that private creditors should reduce their expectations by 40 percent (Vorobiov 2015). Led by the largest private creditor, Franklin Templeton, a giant bond fund, creditors proposed an extension of debt repayments to 2019. They claimed that Ukraine could dig into gold and foreign currency reserves and return more quickly to commercial credit markets and a faster recovery. By summer 2015, the special committee of creditors agreed to write-off 20 percent of Ukraine's $18 billion in bonds held by the foreign lenders and push payment deadlines on most of the rest to 2019. The agreement provided Ukraine with $3.6 billion in debt relief to help close its more than $15 billion funding gap for the national budget. The debt load was 79 percent in February 2016.

Russia would not provide relief (Williams 2015a). Instead it proposed that Ukraine should pay down its eurobond debt in installments of $1 billion, starting in 2016. Ukraine declined, demanding that Russia accept the same terms as other bondholders, as though Ukraine's debt to Russia was a "private" debt similar to that of the commercial bondholders who had accepted a 20 percent haircut, rather than a "sovereign" debt with special protections. In December, the IMF appeared to endorse Ukraine's refusal by altering the IMF rule that if a nation is in financial arrears to any government (i.e., shouldering a "public" loan), it cannot qualify for an IMF loan. A potential split within the IMF board as to whether the organization was an instrument of Western hegemony was inflated by growing alarm as to the competence of Ukrainian leadership – in disarray over whether to maintain support for IMF-imposed "austerity" politics. Ukraine's debt to Russia was recognized as "public" (Mercouris 2015a) and Ukraine was expected to make "good-faith efforts" to restructure the debt. In practice, Ukraine defaulted on the debt in December 2015 and Russia moved to sue Ukraine for recovery.

It seemed likely that the IMF would continue lending to Ukraine regardless, despite IMF admonishments in February 2016 that corruption would have to stop if it was to hand over the next $40 billion tranche. Since only the very naive could have imagined an end to corruption, the IMF was more likely warning Rada politicians to stick to austerity politics. Its effective lending policy seemed to enforce only those debts owed in US dollars to US allies. The IMF had broken several rules for Ukraine's benefit, advancing loans to a country at war that arguably had little chance of repaying them. Civil war in the Donbass damaged Ukraine's productive capacity, export ability and capacity for repaying the IMF (Hudson 2015b). The IMF's initial December decision effectively undercut Russia's influence by detracting from its power as creditor to enforce its will. As argued by Hudson, the issue was exploited by the US-dominated international financial regulatory structure to impede the ability of Russia and China to make loans outside the dollar system to Third World countries that would be guaranteed by the West's regulators (Hudson 2015b).

## Energy issues

Austerity politics in the midst of declining energy demand expedited closure of three mines in Eastern Ukraine (Kiev controlled one-third of the nation's coal mines, most of them in the Donbass) and brought miners out on protest in Kiev in April 2015. But coal was not the key energy issue. A proposed new law would allow foreign investors to lease up to 49 percent of Ukraine's transit pipelines and underground gas storage facilities. This would enable the USA and its allies to control the flow of gas from Russia to the EU, set prices, and make sure that transactions continued to be denominated in dollars (Whitney 2014h).

Russia's aim was independence of Ukraine for transit of Gazprom gas and oil by 2020, mainly by developing two new routes: Nord Stream and South Stream (later converted to Turkish Stream). Either or both of these could deprive Ukraine of transit revenues and weaken its hand in negotiations to buy Russian gas. Ukraine by late 2015 appeared to celebrate anything, even if suicidal, that would detach its economy from Russia's, regardless of economic impact. By this time, however, Nord Stream looked fragile and South Stream/Turkish Stream had virtually collapsed.

The construction of the two routes would likely cost about $12 billion (Golubkova *et al.* 2015). US pressure on Hungary helped to stall Gazprom's plan to build the South Stream pipeline that would carry gas to southeastern Europe across the Black Sea, avoiding Ukraine. Gazprom looked to replace Hungary by means of partnerships with Slovakia and Bulgaria. This strategy was expected to save the $2 billion a year that Gazprom paid to Ukraine in 2014 (about 25 percent of Gazprom's net income), but would cost $1 billion in fees to Slovakia and Bulgaria. The EU protested the Bulgarian connection on the grounds that it contravened its Third Energy Package which, among other things, prohibited a company from both owning the gas and operating the pipeline.

An agreement in principle with Turkey to substitute Turkish Stream for South Stream (bringing pipelines to the Greco-Turkish border without passing through EU countries), had not advanced far by mid-2015. Greek Public Power Corporation (slated for privatization under terms of an EU aid agreement) signed an agreement with Russia in April for a pipeline to take forty-seven billion cubic feet of Gazprom gas. By promoting gas links with Bulgaria (through the planned Interconnector Greece–Bulgaria or IGR pipeline), Romania and Serbia, Greece was also contributing to a diversification of gas supplies in Europe away from the Russian supply and in favor of Azerbaijan and other Caspian locations (Reuters 2015a).

Deteriorating relations between Russia and Turkey at the end of 2015 indicated potentially fatal problems for the Turkish Stream. Russia formally suspended the project in December 2015. With Russia's military presence and influence in the Middle East likely to grow vis-à-vis Syria, Iraq and Iran, Turkey's motivation to diversify sources quickened. In addition to controversial supplies from oil fields controlled by ISIS, Turkey already received gas from the Kurdish Regional Government in Iraq, and a mooted Trans-Anatolian Natural Gas Pipeline (TANAP) would take natural gas from Azerbaijan, across Turkey, to the EU, while a proposed Trans-Caspian pipeline could bring in gas from Turkmenistan, by 2019 (Tanchum 2015).

By February 2016, there were indications of a more favorable EU attitude to South Stream as the EU assessed the likelihood that over time its demand for Russian gas would increase and that Central European economies in particular needed the cheap gas that Russia could supply (Djurica 2016).

Nord Stream involved several collaborations between Gazprom and European energy giants. After considerable progress from 2011 further extensions were announced in June 2015. Germany's largest oil and gas producer, Wintershall, and Gazprom finalized an asset swap worth billions in October 2015. This provided Wintershall, a full subsidiary of BASF – the world's largest chemical company – access to natural gas fields in Siberia. Gazprom expected more influence in the Western European gas market. Gazprom would take on several joint ventures with Wintershall, enhancing its position in the German gas market plus acquiring a 50 percent share of Wintershall's North Sea operations in Norwegian, British and Danish waters. In return, Wintershall would expand its activities in Siberia, and acquire 25 percent of the gas operations at the Urengoy field. The Austrian oil and gas company OMV planned a similar asset swap.

These deals would strengthen Gazprom in the European energy market, and soften the impact of competition from Algeria, Norway and Qatar. Could they off-set the damage of Western-imposed sanctions that had helped reduce Gazprom's market capitalization from $367 billion in 2008 to $51 billion in August 2015? Amidst falling oil prices and the slowing of deals with a troubled Chinese economy, Gazprom also encountered opposition to Nord Stream from the EU, Ukraine, Poland and the Baltic states (Weiss 2015b).

Two other pipeline developments signaled BRICS-favorable energy reconfigurations. One was IP (originally IPI), an Iran–Pakistan pipeline (that once involved India). Resuscitated by Pakistan on the strength of a Chinese commitment to build the Pakistani route and perhaps eventually to connect to China, it had also attracted Gazprom's interest. The other was TAPI (the Turkmenistan–Afghanistan–Pakistan–India Pipeline), dating back to the Clinton presidency of the 1990s. This would traverse Afghanistan (accounting for fabled US support for the Taliban regime in the 1990s, but was abandoned when the Taliban proved incapable of providing the necessary national security [Boyd-Barrett 2003]). It would interconnect Turkmenistan, Pakistan and India, with possible investments from both China and Russia (despite the ongoing friction between Gazprom and Turkmenistan's gas company) (Escobar 2015c). The breaking ground ceremony for TAPI took place in December 2015 with a view to completion in 2019.

These developments held the theoretical promise of stronger Russian influence in worldwide oil production and distribution. But they occurred in the midst of burgeoning sources of supply. These included Iran (following the lifting of US sanctions in the wake of a deal on Iranian nuclear energy), a thawing Arctic, and shale gas drillers worldwide. New sources contributed to a persistent fall in oil prices which, along with US and EU sanctions had induced recession in Russia. The promise of additional demand for Russian gas from China waned as Chinese demand dipped in 2015, forcing the postponement of the implementation of Russia's planned Power of Siberia pipeline to China.

## Shale and fracking

Kiev's resistance not simply to separation but even to federalization was enhanced by the perceived economic benefits of depopulation. Western energy companies sought opportunities for joining with Ukrainian capitalists in fracking enterprises, notably in areas controlled by ethnic Russians (many living, in effect, on a huge shale-gas field [Zuesse 2014b]). The US Energy Information Administration estimated that Ukraine had up to 1.2 trillion cubic meters of gas trapped in its shale rocks, constituting Europe's third largest deposit after France and Norway (Bauerova and Carroll 2014). Ninety percent of Ukraine's oil and gas was located in Eastern Ukraine. (Durden 2014b). The development of shale gas in Ukraine would speed up Ukraine's, and Europe's independence of Russia for energy supplies.

Kiev stopped paying pensions or providing social and financial services to Eastern Ukraine, all but guaranteeing the continuation of the separatist struggle while simultaneously forcing the population out of fracking territory. Global gas price declines and local instability temporarily staunched these ambitions by December 2014. The total number of refugees by mid-2015 reached 1,382,000, at least two-thirds of whom were pensioners or children (UNHCR report July 2015 cited by Stanek 2015). The care of desperate refugees, mainly from eastern regions, was left to volunteer organizations. Most chose to stay in Ukraine; of those fleeing abroad, the majority went to Russia. In April 2015, well after the Minsk 2 ceasefire agreement (see below), President Putin accused Kiev of violating its obligations under Minsk 2 by economically blockading pro-Russian regions (DW 2015a).

Media entities controlled by the energy industry were supplying propaganda for energy interests. Since 2013, when Yanukovych's son Alexander acquired the country's second largest media group, United Media Holding, all of Ukraine's mainstream media were owned by oligarchs. The oligarch and Governor Ihor Kolomoiski (appointed by Kiev to the governorship of Dnipropetrovsk Oblast in spring 2014 and then dismissed in spring 2015) owned the 1+1 Media group. This controlled Ukraine's most widely watched private television channels and the news agency, UNIAN. Through Burisma Holdings and its owner Privat Group – Kolomoiski had acquired it from Mykola Zlochevsky – Kolomoiski owned the biggest share in fracking rights. In 2014, he invited Joe Biden's son, Hunter Biden, to join the board of Burisma. Potentially, the Biden family could become billionaires. Additionally, Kolomoiski owned PrivatBank, which held the bank account for Hromadske TV. Burisma had enough influence over Kiev politics to make prospective gas-fracking lands a military objective. Ukrainian troops helped install shale gas production equipment near the east Ukrainian town of Slavyansk, which Ukraine had bombed and shelled for three months (Hudson 2014a). Civilians protected by the Ukrainian army were installing drilling rigs while the military provided security.

Oligarchs Kolomoiski and Yeremeyev captured Ukraine's lucrative gasoline and petrochemicals markets, Kolomoiski controlling some 1,500 petrol stations and Yeremeyev 500. The USA appreciated Kolomoiski's help in the Donbass where he financed at least two private militia battalions in the fight against separatists. But the oligarch lost kudos when the misdeeds of PrivatBank surfaced. US pressure likely dissuaded him from contesting his dismissal by Poroshenko from governorship of Dnipropetrovsk. His US visa was reinstated (earlier withdrawn when Kolomoiski had threatened Poroshenko, forcibly seizing State oil and oil pipeline companies) to enable his flight to the USA (Cockburn 2015).

## Gazprom and Naftogaz

Amidst price disputes and unpaid bills, the Russian gas exporter OAO Gazprom (GAZP) halted deliveries to Ukraine's state-owned gas company Naftogaz from June to November 2014. Gazprom also reduced supplies to neighboring Hungary and Slovakia to discourage them from sending fuel back to Ukraine. Russia had shut off gas on two previous occasions, ultimately, in 2009, offering Ukraine a discount worth billions. But heavily subsidized domestic gas prices were still lower than the import price. Ukraine's reputation for wasteful energy consumption was deserved. It consumed fifteen times more energy per dollar of GDP than Britain (Sakwa 2015: 73).

Gazprom and Naftogaz relations were strained throughout 2015. This saga underlined Ukraine's interest in establishing alternative sources of supply (assuming it could afford to produce or purchase gas in the first place) and Russia's interest (and, if only logically, Europe's) in establishing delivery routes independent of Ukraine. This was what Gazprom was attempting to accomplish under the Black Sea in partnership with Greece and Turkey (Herszenhorn 2015a;

Reuters 2015k). For the moment though, the prospects for Gazprom looked stony.

Whereas Ukraine imported $4.6 billion worth of gas in 2014, this had been $10.8 billion in 2013 (Timtchenko and Gordiienko 2015). Ukraine's prime minister announced a three-fold hike in the price of gas which he conceded had been much too cheap previously, along with an extension of subsidies for over 2.7 million low-income families. He claimed that 70 percent of Ukraine's gas now came from the West and that Ukraine could be self-sufficient in gas by 2025 (Andersen 2015a; Lough 2015). In October 2014, Ukraine had paid Gazprom $3.1 billion (of a $5 billion total) owed by Naftogaz. From December, Gazprom agreed to supply Ukraine only if it paid in advance. Gazprom reduced supplies again in January 2015, alleging that Ukraine was helping itself to oil from pipelines running through Ukraine destined for European customers. Gazprom told its European customers they would have to connect to the Russian supply on the Turkish–Greek border once the development of South Stream or Turkish Stream was under way. The likelihood of successful completion seemed distant in December 2015 when relations between Russia and Turkey soured following Turkey's shooting down of a Russian jet fighter plane over a Turkish-controlled area of Northern Syria. The USA's and NATO's interest in stalling Russia's Turkish Stream energy ambition helped to explain their complicity with Turkey's action.

In February, Gazprom threatened to stop supplies altogether when the amount of pre-paid gas almost ran dry. Ukraine contested this. The issue turned on Gazprom's insistence that the amount should include gas that Gazprom was supplying directly to separatist areas of Eastern Ukraine which Naftogaz had stopped supplying. Within days of Ukraine using up its prepaid supply, Russia conceded that it would no longer count supplies to Eastern Ukraine, guaranteeing supply up to the end of March, when further discussions on the matter of Eastern Ukraine would take place. The total size of Gazprom's claim against Naftogaz had risen to $29.2 billion by the late summer of 2015 as indicated in a counter-defense statement contesting a Naftogaz claim against Gazprom to the Stockholm Arbitration Court. The continued supply of Gazprom gas in time for 2015–2016, amidst indications that Ukraine had stockpiled enough gas for the winter, was enabled by a $1 billion loan advanced principally from the World Bank and the European Investment Bank. This was terminated in November by Ukraine's claim not to need further supply and Russia's complaint that Ukraine was still behind on its payments (AFP 2015c): Naftogaz had purchased Russian gas from Slovakia. When in January Gazprom presented Naftogaz with a bill for $2.5 billion that it said was owed because Kiev was obliged to purchase a minimum annual quantity of gas, Ukraine's anti-monopoly committee retaliated with a demand for $3.3 billion as a penalty for "abusing its dominant position in the Ukrainian gas transit market."

While WMM demonized Gazprom, less attention was given the development activities of Western oil companies in Ukraine and Crimea through 2014. Western energy companies seriously underestimated the decline in the health of Ukraine's economy – the hryvnia plunged by up to 70 percent against the dollar between early 2014 and early 2015. Additionally, there was a steep decline in oil prices. From a consumer point of view, this offset the progressive elimination of price controls on gas in Ukraine but did not augur well for the economy. Notwithstanding the advantage to them of a declining local currency Chevron, Eni, Exxon Mobil, JKX Oil and Gas, Royal Dutch Shell and others pulled out from agreements for exploring fields in both western and eastern provinces of Ukraine. The reasons cited by analysts and the oil companies included the deteriorating state of the economy, the Ukrainian government's reluctance to modify tax rules in their favor and restrictions of sales to industrial customers. Russia's annexation of Crimea appropriated potentially gas-rich offshore fields in the Black Sea, discouraging explorers like Exxon Mobil Corp. (XOM) and Eni SpA (ENI) (Bauerova and Carroll 2014).

As 2014 progressed, it became more obvious that Ukraine's national economy was extremely frail, contracting by 7 percent that year. A further contraction of between 5.5–12 percent was predicted for 2015. Ukraine was one of only two post-Soviet countries whose GDP had not yet recovered to its 1991 level (Poland's had increased threefold), even though Ukraine's population fell by 10 million to 44 million in that period. One in three of Ukraine's population in early 2014 lived under the poverty line and unemployment was 9.3 percent, yet there were 3–5 million guest workers. The average monthly wage of less than $200, and declining, was four times lower than in Poland, and nine times lower than in Moscow (Sakwa 2015: 73). The shadow economy accounted for 44 percent of output. In 2012, exports were two-thirds of those of neighboring Slovakia, a country with a tenth of Ukraine's population (Sakwa 2015: 72–73). Inflation rose from 1.2 percent early in 2014 to 29 percent in January 2015, reaching 43 percent by December 2015. Foreign reserves dropped by 61 percent in the four months to January 2015 to approximately $6.4 billion (enough for five weeks of imports [Schneider 2015]). Gold had disappeared (taken to the USA) amidst massive outflows of capital. Controls were imposed on purchases of foreign currencies and the interbank currency market. Year on year, food prices among producers rose by 57.1 percent (with grains and vegetables rising 91 percent) while consumption fell by 40 percent (Durden 2015b).

The 2004 color revolution had severed much of Ukraine's economy from Russia (largely accounting for the country's economic stagnation in the following years). From 2004, Russia and the EU as a whole held roughly equal shares of trade with Ukraine – a third each. Russia remained Ukraine's largest national trading partner in 2015 but, whereas trade between the two countries accounted for 14 percent of Ukraine's total foreign trade turnover in the first quarter of 2015, this was a decline from 27 percent in 2013. Further declines between Russia and Ukraine were inevitable given a virtual trade war from early 2016 when Ukraine, after a year's delay, accessed the EU tariff-free zone while exiting Russia's. Russia was concerned that Ukraine's products mainly failed to meet international quality and safety standards and would otherwise flood Russia alongside European products. By this time Ukraine's trade with the EU was two to three times the volume of its trade with Russia. China was the third largest partner.

## Foreign management interventions

Late in 2014, the West sought to guarantee its interests in IMF-imposed austerity by pressurizing Ukraine into an extraordinary measure for any "sovereign" country – introducing into its ranks and providing citizenship for selected foreigners (several would not survive the 2016 Cabinet reshuffle). These included:

*1 Natalie Jaresko*

A Chicago-born Ukrainian American, former US diplomat, who ran the private equity fund Horizon Capital, and now became Ukraine's finance minister. Jaresko had previously run the taxpayer-funded Western NIS Enterprise Fund (WNISEF) founded in the 1990s to jump-start investment economies for Ukraine and Moldova (Parry 2015k, 2015c). Limited to a compensation cap of $150,000, Jaresko shifted management of WNISEF to Horizon Capital, enabling her to secure lucrative bonuses when selling investments (as much as $1.77m in 2013 alone, even as WNISEF itself was making losses). After one year as finance minister, Jaresko had still not renounced her US citizenship – a requirement for Ukrainian citizenship – although she had a year left before needing to provide evidence of renunciation of her US citizenship.

*2 Jaanika Menilo*

The Estonian Jaanika Menilo, a Jaresko associate, was put in charge of attracting foreign investment.

*3 Aivaras Abromavicius*

A Lithuanian investment banker who was previously a partner in the East Capital venture firm, and was appointed Economy Minister. Along with the anti-corruption reformers Oleksiy Pavlenko, Minister of Agriculture, Andrei Pivovarsky, Minister of Infrastructure and Alexander Kvitashvili (see 4), Minister of Health, Abromavicius tendered his resignation in early 2016.

*4 Alexander Kvitashvili*

Formerly Georgia's Health Minister and also a Labor Minister, he now became the Health Minister for Ukraine – a country whose healthcare system has been described as being "in danger of collapse:"

> Soviet-era hospitals across the country of 42 million are crumbling, underpayment of medical staff continues to foster a system of bribe-taking and Ukraine's poor vaccination rate has placed it on a blacklist alongside some of the world's poorest countries, including South Sudan.
>
> (Prentice 2015)

*5 Mikheil Saaskashvili*

Saaskashvili, former president of Georgia – where he was now wanted on charges – was appointed by Poroshenko as the Governor of Odessa.

Their arrival coincided with the establishment of an Information Ministry, and with assurances from deputy prime minister for infrastructure and ecology, Valeriy Voshchevskiy, of the privatization of state holdings such as railway and road (Krieger 2014). The economy continued its decline in 2015 due to fighting in the key industrial regions, and the cutting off of key Ukrainian export markets in Russia. Interest rates were raised by 5 points to 19.5 percent, in an attempt to halt the hryvnia's 46 percent and, later, 70 percent slide against the US dollar since the coup of 2014 (Lantier 2015c). By the summer of 2015 the currency was down 28 percent over the previous year.

Yanukovych (described by Sakwa as a "competent representative of the bureaucratic-oligarchic centrist trend in Ukrainian politics" [Sakwa 2015: 72]) tried to abandon the 2013 EU/IMF deal because it entailed a 50 percent cut in gas subsidies, not to mention cuts in pensions and government employment, and the privatization of government assets. Most of the $15 billion of IMF money would be used to: (1) pay back to the IMF and Western banks the principal and interest on previous loans (about $2 billion a year); (2) finance future exports from the Ukraine, currently running a $20 billion a year trade deficit; and (3) help Ukraine's central bank prop up the hryvnia. There would be little to no positive impact. The "winners" would include:

> western European banks who continue to receive principal and interest payments from the IMF that otherwise would have defaulted; global currency speculators who would be able to sell Ukrainian currency to the Ukrainian central bank at a subsidized price; Ukrainian companies that would be given export credits to continue selling to western Europe; and the western Europe companies that import the Ukrainian exports at a more attractive price.
>
> (Rasmus 2014: 123)

Much the same pertained when the IMF agreed to new loan conditions totaling $17.5 billion in March 2015. Initial payments were scheduled for key creditors, including US financial firms such as PIMCO, BlackRock and Franklin Templeton investments.

The decline flattened modestly by the second half of 2015. Moody's ratings agency upgraded Ukraine's sovereign rating to "stable." Standard and Poors affirmed a 'B-/B' long- and short-term local and foreign sovereign credit rating. The rating acknowledged a restructured debt, passage of major reforms and austerity policies. Significant corruption remained along with unrest in the East and instability in government.

Government statistics showed that in the first eleven months of 2015, the deficit was fourteen times lower than in the previous year, national budget revenues were up 50.4 percent and expenditures up 27.1 percent (UNIAN 2015l). Ukraine's debt had fallen in less than two years to $66 billion from $72 billion. Economic decline was down to 7 percent in the third quarter. Industrial production slowed to 14.4 percent in the period from January to November 2015. The sales of industrial goods in January–August 2015 were 22.3 percent up year-over-year (UNIAN 2015m) and in 2014 they were 6.73 percent up on 2013. The sectors demonstrating production growth from August 2014 to August 2015, included production of coke and fuel (24.7 percent), the manufacture of computers, electronic and optical equipment (22.3 percent), the mining of metal ores (6.7 percent) and coal production (1 percent). Ukraine's national budget from January through to September 2015 yielded a UAH 8.8 billion surplus in sharp contrast to a UAH 40.1 billion deficit recorded over the same period in 2014. Ukraine's foreign trade surplus increased by 28 percent in the first nine months of 2013. The Agriculture Ministry forecast an increase in grain exports for 2015/6 to 36.6m tons (out of a total harvested of 59.2m tons) (Reuters 2015h). Natural gas production increased by 2 percent from January to November 2015. In the third quarter, Rada secured a constitutional majority for a government-prepared bill to increase wages and pensions by 13 percent and 19 percent for wages paid to highly skilled personnel and 13 percent for pensions (UNIAN 2015n).

Yet year-over-year inflation was 43.3 percent in December 2015 as compared with 24.9 percent in 2014 – the highest level in twenty years (UNIAN 2016a). The Ukraine Statistics Service reported an increase in average full-time wages of only 3.3 percent between August 2014 and September 2015 to UAH 4343 (or $190), although the number of full-time employees remained unchanged at 8 million. The highest salaries were reported in air transport at UAH 19600 or $856 (UNIAN 2015n). In the second quarter of 2015, GDP fell by 14.6 percent year-on-year. The banking system remained fragile: total losses were at $4.2 billion (4.4 percent of estimated GDP in 2015) in the first seven months of 2015.

The production of thermal and coking coal halved in the period January to September 2015 year-on-year, given continuing unrest in the Donbass. Production was down from 53.4 million tons to 29.6 million, a further decline from the 65 million level at the end of 2013 (Reuters 2015i; UNIAN 2015a). Ukraine's gross foreign debt in the second quarter of 2015 grew by 0.8 percent, or $1.007 billion, from the first quarter of 2015, to a total of $126.976 billion. The country's gross foreign debt to GDP grew to 122.8 percent from 110.5 percent of GDP in the first quarter (Reuters 2015h). Ukraine received a mere 0.8 percent of planned receipts from privatization in 2015. Ukraine–Russia trade turnover halved in the first nine months of 2015. Yet Russia still remained Ukraine's largest national trading partner, with a 12.8 percent share in Ukraine's exports and providing 20 percent of the country's imports. Ukraine's next most important export markets were Turkey, China and Germany.

WMM coverage failed to give appropriate salience to the long-standing fragility of Ukraine's economy, the extent to which it had been intertwined with that of Russia, or the consequences of artificially slashing that relationship. In place of blaming Ukraine's oligarchs and their hold over the country's political parties for the country's perilous economy, WMM – with little to no evidence – preferred to blame Russia for all Ukraine's ills. The importance of Gazprom as source of energy for Ukraine was generally portrayed as something sinister, rather than as a form of Russian subsidization of Ukraine's economy connected to Gazprom's dependence on Ukraine as a transit route to Gazprom customers in Europe. The IMF loan package was portrayed as a miracle god-send that would rescue Ukraine from evil Russia, whereas in fact the alternative Russian deal on offer would have been much less punitive, inflicting far less damage, if any. Rarely did WMM consider that the US/NATO/EU conspiracy to entice Ukraine into the Western orbit through the 2014 coup almost immediately precipitated an internal crisis, a civil war, the loss of Crimea, and the de facto ruination of Ukraine's major industrial region in the East. Such destabilization was reminiscent of previous Western destabilization campaigns in Yugoslavia, Iraq, Libya and Syria. For Ukraine's people and economy, the consequences were catastrophic. Other than the mindless targeting of Russia as culprit, rarely did WMM investigate the parties most likely to benefit from these acts, including Western agribusiness, energy giants and other Western investors eager to cash in on the cheap privatized assets of Ukraine's former State economic apparatus.

# 6 Crimea, Odessa and Eastern Ukraine (Donbass)

## Crimea

When the coup attempt looked fragile early on, WMM speculated that Kiev could sever Western Ukraine as a separate entity. Concerns that later caused such anguish when Crimea and the Donbass fought for separation or autonomy were not in evidence (Nazemroaya 2014a). The *New York Times* instead smeared the movement for Crimean secession (Parry 2014k). The *New York Times* correspondent C.J. Chivers co-authored a dispatch with Patrick Reevel titled "Pressure and Intimidation Grip Crimea," with the subtitle, "Russia Moves Swiftly to Stifle Dissent Ahead of Secession Vote." The journalists alleged Russian intimidation, military occupation, and electoral manipulation ahead of the referendum on March 16 (Chivers and Reevel 2014). In various reports, the *New York Times* and other WMM suggested that Russia had "invaded" Crimea. They attached little importance to the fact that Moscow already had 16,000 troops stationed in Crimea under agreement with Ukraine allowing Russia to maintain up to 25,000 troops to protect its historic naval base at Sebastopol. Russian troops did support the Supreme Council of Crimea as preparations were under way for a referendum that soon demonstrated overwhelming public support for secession (Chivers and Reevel 2014; Parry 2014l). This predictable result owed little or nothing to Russian "occupation" or electoral "manipulation," and much to long-standing ambivalence towards Kiev among a strong majority of Crimeans, especially among ethnic Russians.

Greta Uehling, whose perspective was influenced by her studies of the Crimean Tatars, offered a less fulsome account claiming, without sources, that Russian troops "took over" the Supreme Council, installing pro-Russian Prime Minister Sergei Aksyono whom she asserts had criminal ties, and that the Council *then* held a referendum (Uehling 2015). The criminal ties are disputed (*Der Spiegel* reported mafia connections, but based on German intelligence documents; cf. Bidder 2014). Ukraine condemned the Russian annexation to the UN General Assembly. Its resolution called on member countries not to recognize the redrawing of Ukraine's borders. This was passed on March 26, 2015 by 100 votes to eleven, with fifty-eight abstentions. The abstentions included Bolivia, China, India, Israel, Lebanon and South Africa. UN resolutions of this kind lack enforcement power and are frequently ignored by major powers, including the USA which routinely ignores resolutions condemning Israel's occupation of Palestine.

Two weeks before the referendum, Vladimir Putin acknowledged that Russian troops present in Crimea would resist its occupation by Ukraine's coup regime, in support of the wishes of the Crimean people. WMM negativity persisted into early 2015 when the *New York Times* reported a claim from the opposition Moscow newspaper *Novaya Gazeta* that the conservative Russian oligarch Konstantin V. Malofeev provided the Kremlin with advance warning of the Yanukovych government collapse in February 2014 and counseled Russia to exploit the ensuing chaos by annexing Crimea and South Eastern Ukraine while offering in justification the EU's own rules on self-determination. The *New York Times* report had no evidence of such a memo, nor evidence that it had been applied for policy purposes. Its existence was denied by both the Kremlin spokesman and by Malofeev. Its contents did not square well with Russia's actual behavior: (1) caution on the question of Crimea, where it acted only until after a Crimean referendum confirmed broad popular support for annexation; and (2) refusal to annex other regions discontented with the coup regime, despite persistent local appeals to Russia to do so (MacFarquhar 2015).

Evidence of a strong pro-Russian tendency (including also a separatist tendency) in Crimea, where Russia's presence dated to the early 1700s was abundant. Crimea formally became a part of Russia in 1784. The province had previously formed part of Crimean Khanata of the Tatars, sometimes said by Tatars today to have been one of the strongest and most independent powers of Eastern Europe even though it was absorbed within the Ottoman Empire. This prior history sustained the ethnic nationalism of the anti-Russian Crimean Tatars, whose hostility dated to Stalin's wholesale deportation of the Tatars to Central Asia in 1946 (40 percent died in the journey). This was said to be punishment for alleged Tatar collaboration with Germany in World War II. There were no Tatars in Crimea in 1959, and they constituted only 12 percent (200,000) of the population in 2001. Of these, perhaps 20 percent considered themselves aligned to the Tatar's Mejlis parliament which claimed to speak on behalf of all Tatars. Tatars were only able to repatriate in substantial numbers after 1991 and in 2014 represented a struggling Crimean minority. During 2015, Tatars allied with Ukrainian nationalists in support of actions to sabotage energy, trade and transportation links between Ukraine and Crimea. Ukrainians comprised 24 percent of the population of Crimea in 2001.

Fifty-eight percent of Crimea's two million people in 2001 claimed to be Russians and 77 percent were registered as native Russian-speakers (Sakwa 2015: 13; Lubin 2014). An absolute majority of Crimeans (97 percent) spoke Russian as their main language, according to Kiev's International Institute of Sociology (RT 2014a). Ethnic Russian Crimeans, particularly, reacted passionately against the revocation by Kiev's new authorities of a law that legalized the regional use of minority languages, including Russian. This allowed Ukraine's Russian-speaking regions to use Russian in official business and education. The new government also proposed to prohibit members of the former regime from occupying official posts.

Ethnic Russians in Crimea long resented Soviet Premier Khrushchev's arbitrary 1954 transfer of their region to Ukraine. A strong pro-Russian separatist movement emerged in the early years of Ukrainian independence. Overtly pro-Russian separatist organizations, including the Communist Party of Crimea and the Party of Regions, did not do well in the mid-1990s elections, but separatism remained a significant possibility. Seventy-three percent of decided Crimeans in 2008 backed secession of Crimea from Ukraine and joining with Russia. The support for separatism increased after EuroMaidan (Katchanovski 2015a: 83–86). A 2015 poll indicated that 82 percent of Crimean people fully supported their region's inclusion in Russia, while only 4 percent spoke out against it [Kelly 2015]).

In March 2014, the Crimean government invited citizens to choose between staying part of Ukraine – but with greater autonomy – or joining Russia. Voter turnout was reportedly 80 percent, and the vote to join Russia was 97 percent [Roberts 2014b]). A report of the Russian Presidential Council for Civil Society and Human Rights later estimated the turnout to have been only between 30–50 percent, of whom 50–60 percent voted for unification with Russia, with a higher turnout of 50–80 percent in Sebastopol where the overwhelming majority voted in favor (Sakwa 2015: 104). Sakwa concluded that even in perfect conditions, a majority in Crimea would have voted for union with Russia and that in Sebastopol the favorable vote would have been overwhelming. In a Pew Research Center survey in April 2014, 91 percent of respondents in Crimea stated that the referendum was free and fair (Pew Research Center quoted by Katchanovski 2015a: 86). Large sections of Ukrainian military, security service and police personnel in Crimea switched their allegiance to the separatists and then to Russia. In the September 2014 elections, the United Russia party of President Putin won 71 percent of the votes in Crimea, although some opposition parties were constrained in how they were able to function at this time (Katchanovski 2015a: 86–87). Russia's *Kommersant* newspaper reported the results of a survey of Crimeans conducted in November 13–16 by the Levada Center, an independent Russian pollster: it showed that 87 percent of respondents believed Crimea should be part of Russia. Only 3 percent thought the peninsula should be part of Ukraine (Annis 2015). These findings were consistent with the results published by several Western-funded polls in late 2014.

The autonomous Rada of Crimea had long been at odds with Kiev. A movement to unify Russia and Crimea was of long standing. Crimea's Rada passed anti-NATO legislation in 2006, banning NATO forces from entering Crimean territory (Nazemroaya 2014a). Writing for *Forbes* in March 2015, Kenneth Rapoza cited several surveys demonstrating that Crimeans of all stripes – ethnic Russians, Ukrainians, Tartars – overwhelmingly wished to belong to Russia rather than be returned to Ukraine. The surveys included a June 2014 Gallup poll sponsored by the US government's Broadcasting Board of Governors, and a February 2015 poll by Germany's GfK (Rapoza 2015; Parry 2015b). One strong practical reason for loyalty to Russia was that Russian pensions were three times higher than Ukrainian ones, and that Ukrainian pensions were being slashed in concession to IMF demands for austerity.

A rare WMM attempt at full historical context was offered by the British newspaper, *Mail on Sunday*, almost a year after Crimea voted for annexation with Russia (Hitchens 2015). In December 1991, the UN formally recognized Russia as the "continuer state," which technically meant that everything that belonged to the Soviet Union came under Russian jurisdiction, including Sebastopol – "an object of all-union significance." In January 1991, before Ukraine's independence in August of that year, the government of Crimea held a referendum on Crimean autonomy. The vote in favor of autonomy was 93 percent of an 80 percent turnout. In March, 87 percent of Crimean voters voted to stay in the Soviet Union: they anticipated this would afford them greater independence than if they were incorporated within Ukraine. Shortly before the final collapse of the Soviet Union, Crimea's parliament approved by 153 to 3 a measure that would enable the region to hold a referendum on its political future but the newly independent government of Ukraine, once in power, prohibited the referendum even though it had been requested by 246,000 of Crimea's 2.5 million people.

Moscow had done nothing to prevent Ukraine from declaring independence, nor protested Ukraine's refusal to allow Crimea a referendum vote. President Yeltsin indicated that Russia would not let republics with large Russian populations secede from the newly emerging states of the former Soviet Union, but later backed down. When signatures for a referendum were being collected in 1992, Ukraine offered more autonomy for Crimea. Crimea's parliament voted 118 to 28 for secession on May 5, subject to confirmation by referendum. This vote was reversed the following day amidst threats of bloodshed and direct presidential rule from Ukraine, and the carrot of greater autonomy. Ukraine declared the May 5 vote unconstitutional and plans for a referendum were canceled.

WMM understated how events in Crimea were an inevitable response to Western meddling that had precipitated the collapse of the Yanukovych regime on February 22 (Smith 2014). Khrushchev's 1954 gift of Crimea to Ukraine did not include Sebastopol, site of Russia's Black Sea fleet. When Ukraine became independent following the Soviet Union's collapse, territories that had previously been Russian but then appended to Ukraine by Soviet rulers were retained by Ukraine, under Washington pressure. In compensation, Russia was given a fifty-year lease on Sebastopol.

Several legal justifications supported the Crimean referendum for annexation with Russia (Sakwa 2015: 108–110). Secession may have been unconstitutional under the constitution in force up to February 2014, but the February 2014 coup rendered that constitution null and void. Procedurally, the 1954 transfer of Crimea was not correct. Additionally, the port of Sebastopol, as an "object of all-union significance," remained under Moscow's control even after the dissolution of the Soviet Union. The principle of inviolability of borders, invoked with respect to the Russo-Ukrainian treaty of 1990, applied to administrative divisions within a single Soviet state, not to independent countries. The annexation averted attacks against ethnic Russians of the peninsula that they had good reason to fear in the wake of the coup in Kiev and subsequent massacres in Odessa and the Donbass and in the context of long-standing dissatisfaction in Crimea with Ukrainian attempts to impose cultural hegemony. The rights of peoples to self-determination, finally, is a cardinal principle of modern international law.

There were certainly problems with the circumstances in which the referendum was conducted: the presence of armed troops, the hurry with which the referendum was conducted, the absence of independent international observers and the lack of transparency in counting procedures. But if constitutional behavior had broken down in Kiev, there was no reason to preserve it in Crimea. In addition, the West had created numerous precedents that in a less partial environment would have required their support for the Crimean declaration: for example, Western support for Kosovo's unilateral declaration of independence from Serbia in 2008, which was declared without even first staging a referendum.

WMM had shown no insuperable doubts in accepting the basic legitimacy of an Iraqi election in 2005 which was conducted under extremely violent US–UK military occupation, press censorship and control, and vote rigging. Nor had WMM gagged in the face of Libya's elections in 2012, conducted amidst the chaos and societal fragmentation that followed NATO intervention. Nor, again, in face of the Afghan presidential elections of 2014 carried out under equally non-propitious circumstances. Nonetheless WMM were adamant in their refusal to acknowledge the legitimacy of Crimea's referendum which, they insisted, was conducted "at [Russian] gunpoint" (Edwards 2014) or because it was "under military occupation." They refused to countenance comparisons between Russian action in Crimea and Western support for separatism elsewhere. In 1999, as we have seen, NATO had fought a war against what remained of the Federal Republic of Yugoslavia with the goal of militarily occupying Kosovo province and prying Kosovo away from Yugoslavia. No referendum on Kosovo's separation from Yugoslavia was held (although, to be sure, the measure was broadly popular in Kosovo). By contrast, in their coverage of Ukraine, the word "unelected" was rarely attributed to the interim Kiev rulers, and the word "coup" was almost always used in quotes.

The editors of the media monitoring website Media Lens identified the BBC's pro-West stance, noting that BBC News "sticks to a propaganda framework which reflects the values and priorities of the UK government and wider western powers" (Cromwell and Edwards 2014). At its least pernicious, this took the form of headlining complaints from the UK Foreign Secretary William Hague or the US Secretary of State John Kerry that the "territorial integrity" of the Ukraine had been violated – a view conveniently and hyper-hypocritically unembellished by any reference to the West's invasions, occupations and/or destabilizations of Iraq, Afghanistan or Libya, an omission that Cromwell and Edwards considered was bordering on "contempt for public memory and understanding of recent historical events." The BBC, they believed, acted as though it was mandatory "to portray Russia largely as a threatening dangerous power in a manner that the broadcaster does not do with the UK, the United States or NATO" (Cromwell and Edwards 2014).

Equally missing from the repertoire of Ukraine coverage was the thorny issue of the approval by the UN's 2010 International Court of Justice for the separation of Kosovo from Serbia. Kosovo had been a constituent part of Serbia for centuries and was internationally recognized as such from 1912. Because Kosovo had never been a Yugoslav republic, unlike the six actual Yugoslav republics, it had no right under international law to become an independent state (Kuzio 2015: 115). It is useful at this point to bear in mind the claim of French Brigadier-General and famous geo-strategist Pierre Marie Gallois, that the balkanization of Yugoslavia was planned much in advance by Germany and the United States in the 1970s, rather than at its actual occurrence in the 1990s (Shukla 2015). The USA was keen on this because it wanted NATO to continue in a post-Cold War world; it needed Europe to depend on it, militarily and wanted to humiliate Russia, testing its Slavic unity with Serbia. The former CIA agent and author Mike Bauer has claimed in an interview to have been assigned by the agency to missions in the former Yugoslavia whose purpose was to incite hatred between the republics and blame Serbia.

> It is all very clear, the people who incited the war and dictated the terms of the peace now own the companies that exploit various mineral resources and the like! They simply made slaves of you, your people work for nothing and that produce goes to Germany and America [...] they are the winners!
>
> (Cited by Britic 2015)

After the collapse of the Soviet Union in 1991, Western powers backed the separatist bid by Kosovar Albanians, in line with the longer-term Western interest, as we saw above, in dismembering the former Yugoslavia. Previously successful backing for Croatia and Slovenia's unilateral declarations of independence from Yugoslavia in 1991 provided further encouragement. Within months of Kosovo's declaration of independence in February 2008, Georgia attacked the separatist enclave of South Ossetia, provoking a brief war with Russia. The USA adamantly rejected South Ossetia's demand for independence (Hyland and Mitchell 2010).

Certainly there was a heavy price to be paid for Crimean independence. For example, Crimea relied on Ukraine for 85 percent of its water. Ukraine blocked water supplies in 2014, claiming that consumers had not paid their debts. Also, 82 percent of Crimean electricity came from Ukrainian plants (a dependence that was being rapidly redressed in 2016). There were issues as to which country Crimean enterprises needed to pay taxes to and the implications for their vulnerability to US and EU sanctions. The infrastructural reintegration of Crimea with Russia would be a lengthy process, with cultural and political challenges, including an increasingly intolerant attitude towards dissenting journalistic and other dissident or potentially dissident voices – albeit arguably on a par with Kiev's attitudes to dissent against the coup regime, as illustrated by the silencing of the popular opposition daily newspaper *Vesti* and the ousting of its editor-in-chief Igor Guzhva (*Russia Insider* 2015). The standard of living, employment and pensions in Crimea were many times superior to those of Ukraine but Uehling's chronicle of a drift towards authoritarianism under Sergey Aksyonov (both prime minister and head of the Supreme Council), forced disappearances, searches and shrinking freedom of the press and speech – and implications of these, in particular, for Ukrainians and Tatars – cannot easily be dismissed (Uehling 2015). The United Nations Human Rights Monitoring Mission in Ukraine (HRMMU) expressed concern in December 2015 about instances of human rights abuses in Crimea where it claimed residents "continued to be affected by broad curtailment of their rights due to the application of a restrictive legal framework imposed upon them by the Russian Federation" (UNIAN 2015q).

Jonathan Cook, a former *Guardian* correspondent, took to task the *Guardian*'s Moscow correspondent, Shaun Walker, for a long article by Walker which omitted mention of Russia's two major concerns namely, Western hypocrisy over Crimea, and the threat posed to Russia by NATO expansionism (Cook 2014). In place of talking about Russia's legitimate concerns about NATO bases encircling Russia, Walker quoted US diplomats to invent something called "Eurasianism" – an alleged Russian ambition to take over Europe, using Ukraine as the launch-pad – for which there was no evidence.

In WMM coverage, Crimea had low visibility after its 2014 annexation until November 2015, and even then coverage was muted about the sabotage of Ukraine's electricity supply to Crimea by Ukrainian and Tatar nationalists. The indecisive WMM reaction was partly due to the accentuation of tensions during November over Russian and Western involvement in the Syrian crisis, and the shooting down of a Russian fighter jet by Turkey. Having called the West's bluff on its seriousness of intent in fighting ISIS (as opposed to backing ISIS or Al Qaeda-linked jihadist rebels against Assad's regime in Syria), Russia initiated its own significant bombing campaign.

The Crimean peninsula was severed from Ukrainian power grids on November 20. Two of four transmission pylons on the administrative border between Ukraine and Crimea were blown up leaving half of Crimea's 2.3 million people without power. The sabotage was variously attributed to Crimean Tatar rebels and/or Ukrainian nationalists, who obstructed repairs in an attempt to secure the release of political prisoners, to make Russia and Crimea pay a hefty price for the annexation, and to bully the Ukrainian President Poroshenko into abandoning his more laissez-faire Crimea policy. Tatar spokesmen complained when Kyiv negotiated to restore electricity. An agreement that the region should be a free-trade zone was suspended in response to the Tatar protests. Human-rights violations were downplayed (Coalson 2015). The saboteurs were successful in eliciting a harder line towards Crimea by Poroshenko. Their action conformed to a broader trend towards a blockade between Ukraine and Russia generally.

Initially the electricity blackout was total in the cities of Feodosia, Yevpatoria, Simferopol (capital) and Sebastopol. It affected the supply of petrol because pumps no longer functioned. Ukraine too was affected: the regions of Kherson and Nikolayev were deprived of electricity; two of Ukraine's nearby nuclear energy stations were endangered. One source claimed that to access and repair the lines, the Kyiv government had to negotiate with Ukrainian nationalists by offering, in concession, a total ban on cargo traffic to and from Crimea, amidst threats to annul contracts for power supply to Crimea. This was a response to Russia's ban on Ukrainian food imports from January 1 – Russia's response to the EU–Ukraine "economic association" scheduled for that date and against the continuation of EU–US–Ukraine sanctions impacting Russia's banking, defense and oil sectors, and blacklisting selected individuals. At the G20 October summit meeting in Turkey, Western leaders had extended their embargo against Crimea and sanctions against Russian political and business leaders by six months. The ban on cargo traffic would add to a food transport blockade of Crimea which Ukrainian nationalists had enforced since September 20, largely ignored by WMM.

Russia supplied emergency electricity and mobilized gas turbine and diesel generators. Mainland Russia started supplying electricity to Crimea in the first days of December, after completing the first power bridge to the peninsula across the Kerch Strait. This would deliver 200 MW of electricity a day from the Krasnodar region in mainland Russia. Another 200 MW would be delivered by a second power cable, starting December 20. These measures addressed 80–90 percent of energy needs. Eastern Crimea was first to receive electricity from Russia, mainly benefiting the cities of Kerch and Feodosia as well as 180 other settlements. Another 400 MW energy bridge was scheduled to open in May 2016 (RT 2015i). In a poll conducted by the All-Russian Public Opinion Research Center (VTsIOM), over 93 percent of respondents voted against signing a new electricity supply contract with Ukraine when the existing contract expired in December. Crimeans preferred the continuing threat of blackouts to renewing an agreement that insisted on describing Crimea as part of the Ukraine. A few days later, further sabotage incapacitated concrete pylons supporting power lines near the village of Bohdanivka in southern Ukraine's Kherson region – the "Kakhovka-Titan" line to Crimea, the only one of four lines to be restored following the explosions of November 22.

Because Ukraine obtained a significant portion of electricity – and coal for its generation – from Russia, the destruction of the pylons indicated a protracted dispute. In a December 2014 agreement, Ukraine had undertaken to sell electricity to Crimea and, in exchange, Russia was to sell electricity to Ukraine (Annis 2015). Ukraine produced over half of its electricity using nuclear power plants. Nineteen nuclear reactors were in operation, with two more under construction. The nuclear fuel for these reactors was being supplied by Russia (Orlov 2015). The danger to nuclear reactors of lax maintenance in an unstable situation of civil war, in the home of Chernobyl, and subject to potential destabilization by power blackouts seemed of little concern to WMM.

## Odessa

When armed protesters occupied government buildings in ten cities of Eastern Ukraine in April 2014, WMM repeatedly cited US officials blaming Russia. One of the insurgents' first acts was to take over television stations and restore Russian broadcasts cut by order of Kiev authorities (Sakwa 2015: 150). Ignored were the protestors' main concerns. These included the Kiev regime's threat to abolish Russian as an official language, even though 30 percent of Ukrainians considered Russian to be their first language and were Russophones. Protestors distrusted the appointment of billionaire "oligarchs" as regional administrators. They rightly worried that a harsh austerity plan would accompany the IMF "aid" now scheduled to replace the less onerous loan previously offered by Moscow.

The *New York Times*' front-page photos attempted to prove that Russia was behind the occupations. Robert Parry told Anton Woronczuk on *Real News Network* that at face value the photos – provided courtesy of the US government – showed fighters alleged to have been in Russia and who were now in Eastern Ukraine. This was hardly proof of Kremlin involvement. A key photo, supposedly taken in Russia, was declared by its photographer to have been shot in Eastern Ukraine. The *New York Times* issued a retraction (Woronczuk 2014). The "newspaper of record" had not undertaken elementary fact-checking of material handed to it by the Kiev regime and the US State Department before plastering its alleged "evidence" on its front page. This raised:

> the question of how the *Times*' photo report was published in the first place. Either the *Times* editors rammed through the piece and its blurry photos without any independent examination, or the editors did check the story, saw it was a grotesque falsification, and published it anyway. In either case, the *Times* functioned not as a legitimate journalistic outlet, but as a propaganda agency of the state.
>
> (Lantier 2014b)

The *Washington Post* reporter Lally Weymouth (2014) referred to "Russians" in occupied buildings, and at another point called the protesters "terrorists." No proof had been presented by the US government that Russians had taken part in the occupations (Johnson 2014). For several weeks WMM spoke of covert Moscow involvement behind Eastern Ukrainian resistance, which Russia media indignantly rebutted. The *New York Times* dispatched the reporters C.J. Chivers and Noah Sneider into Eastern Ukraine. The reporters surprised themselves, their newspaper, and its readers by finding little if any such evidence (Chivers and Sneider 2014; Parry 2014g). Investigating the 12th Company of the People's Militia, Chivers and Sneider found only Ukrainians, even if many had affinities, family connections and other ties to Russia. The Ukrainians had many different views as to what might constitute a good outcome to the conflict. Weaponry in evidence appeared well-worn and dated, not the equipment that one would expect if Russia had been actively intervening. In evidence of local discontent, Anthony Faiola of the *Washington Post* reported on April 16 that the East Ukrainians he had interviewed were fearful of the IMF austerity package to which the new regime was committed (Faiola 2014).

In the tragic Odessa events of May 2, local ethnic Russian or separatist protestors voicing their opposition to the Maidan *coup d'état* were set upon and their tents burned in what appeared to be a well-organized and provocative action. Fascist pro-Kiev paramilitaries from outside of Odessa incorporated into units of the National Guard were implicated in the atrocity. Many were veterans of the Maidan violence; they included police sympathizers (Zuesse 2015b).

The leader was the interim government's minister for security, Andriy Parubiy, the founder of the Social-National Party of Ukraine in 1991, which blended nationalist and Nazi symbology. His men were regularly recruited to undertake the dirty work rejected by the regular Ukrainian military. Many victims were forced to flee into the Trade Unions Building. This was then attacked by grenades and set on fire while attackers sang nationalist songs and shouted "Burn Colorado, burn" the fascist sobriquet for Colorado beetle colors of the anti-Maidan protestors' flag. Graffiti on the burnt building included swastikas and allusions to the fascist Galician SS of World War II – over forty died. Those who escaped were severely beaten. A similar atrocity occurred on May 9, when the same thugs burned down a police station occupied by pro-Russian protestors, including policemen opposed to the coup regime in Kiev. Between seven and twenty were killed. WMM referred to the killers as "volunteers" or "self-defense" forces (Parry 2014h).

Four commissions were established to investigate the Odessa massacre, yet much remains unclear (Sakwa 2015: 98). One analyst speculated:

> The lack of commiseration, empathy, or compassion in relation to the death of "Colorados" in Odessa, "jokes" on the blood of Russian babies on a school charity fair, the openly Russophobe art exhibition titled "Kill a Colorado" that took place in Kiev in December – those are not accidental events. They have been prepared by Ukrainian intelligentsia.
>
> (Pogrebinskiy 2015: 97)

In an otherwise anti-Russian speech in June 2015, the US Ambassador to the UN Samantha Power criticized the post-coup regime – noting that:

> investigations into serious crimes such as the violence in the Maidan and in Odessa [where scores of ethnic Russians were burned alive] have been sluggish, opaque, and marred by serious errors – suggesting not only a lack of competence, but also a lack of will to hold the perpetrators accountable.
>
> (Cited by Parry 2015i)

Writing for alternative media site *Global Research*, Parsons (2014b) observed that:

> neither the President nor Secretary Kerry see any contradiction in their applause for the Kiev government as representing democracy while it was a neo-Nazi coup that ousted a democratically elected President and that while the President and Kerry were "disgusted" by the violence on the streets of Kiev in mid-February … [they] … have no reaction to the deaths in Odessa and other east Ukraine communities.

If the coup was really about democracy, she asked, what was so democratic about the violent attacks on those citizens of Odessa who did not favor the coup? What was so wrong with their protest that it deserved such a vicious response? Protestors had used the same tactic as the coup leaders had employed in Maidan – peaceful occupation of a public building – while insisting on a referendum to consider a more autonomous future.

Despite the high proportion of ethnic Russians among its citizens, Odessa remained under Kyiv control. In June 2015, the Ukrainian President Poroshenko appointed the ex-president of Georgia, Mikheil Saakashvili to the governorship of Odessa. Saakashvili was a neo-conservative who had dragged his country into an unsuccessful war with Russia in 2008. The government that came to power in Tbilisi after defeating Saakashvili's party in 2012 parliamentary elections, accused him of fraud, organization of an assault and abuse of office – charges he denied. Disgraced, he spent the period between leaving Georgia and becoming governor of Odessa in Brooklyn, New York (Parry 2015e). Georgia's request for his extradition from Ukraine was denied on the grounds that the charges were politically inspired. The Georgian President Giorgi Margvelashvili revoked Saakashvili's citizenship in December 2015, on the recommendation of Justice Minister Thea Tsulukiani, given Georgia's ban on dual citizenship and Saakashvili's assumption of Ukrainian citizenship (RFE/RL 2015b).

Another city that witnessed the occupation of buildings in May was Mariupol which, like Odessa, remained unsteadily in Kiev hands. The *New York Times* reported on May 23 that buildings in Mariupol Makeyevka had been "recovered" by steel workers and other employees of Ukraine's richest billionaire, Rinat Akhmetov – one-time enforcer for a Donetsk crime boss who had used extortion and physical violence to seize control of former state-owned property and whose businesses in 2014 accounted for 25 percent of Ukraine's GDP. Other papers, including the *Washington Post* and the *Financial Times* saw no evidence of this. Rather, as Van Auken argued, the *New York Times* had gleaned its information from Akhmetov's press agents and that:

> the occupied administration building in Mariupol was not "recovered" by miners and steelworkers; it was gutted in a brutal assault from the Ukrainian military and Right Sektor fascists in which at least 20 civilians were murdered in a hail of automatic weapons and tank fire.
>
> (Van Auken 2014a)

## Donbass

The Crimean referendum was followed on May 11 by autonomy referenda in the provinces of Eastern Ukraine, although ultimately Russia discouraged calls for the annexation of these provinces (where, Sakwa argues, the prevailing sentiment favored greater autonomy *within* Ukraine – not separatism, nor annexation) and tried to have the referenda postponed. Russia had good reason to tread carefully. The concept, popular in the Donbass, of a "Novorossiya," was associated with the so-called Russian Spring movement that proclaimed not only revolution against Kiev but also against the existing regime in Moscow (Laruelle 2015).

In Donetsk, protesters proclaimed the Donetsk People's Republic on April 7. The Luhansk People's Republic was formed on April 27. In the ensuing May 11 referenda, turnout in both regions was reported to be 75 percent, with 89 percent and 96 percent respectively voting for independence (Sakwa 2015: 154). A month later, on May 24, the two entities established a union they referred to as the Novorossiya Republic.

Neither Kiev nor the West recognized the May 11 ballots as legitimate. WMM routinely declared them as farcical. As Ukraine readied for war in the Donbass, the West and WMM found themselves supporting the deployment of an "army against their own population," argued Escobar (2014c), the very action used by NATO as the pretext for interventions in Iraq, Libya and Syria but apparently acceptable in Ukraine. Richard Sakwa considered that Kiev's ferocious reaction reflected an established prejudice in Western Ukraine that the peoples of the Donbass were not "real Ukrainians" but Russians sent in the 1930s, first to replace those who had died under Stalin's collectivization program and second to industrialize Ukraine. Votes for independence were to be interpreted not as bids to leave Ukraine (since separatist aspirations were not supported by the majority here as they had been in Crimea) but as demands for a pluralist, federalist Ukraine. Sakwa refers to the polls of early 2014 showing that sizable majorities in the west and south supported Ukrainian unity, with small minorities favoring secession or accession to Russia (Sakwa 2015). Mere discussion of federalization was outlawed from mainstream discourse in greater Ukraine (Pogrebinskiy 2015: 97–98).

Challenges to the legitimacy of the autonomy referenda were predictable: rush organization in conditions of civil war and threats from the post-coup regime. Such circumstances were scarcely less dire than the Western-induced post-Gadhafi elections in Libya in 2012 which WMM applauded as fine displays of democracy. Despite the challenges, the turnout in the Donbass was huge and victory was a landslide. The transparency was impressive: "a public vote, in glass ballot boxes, with monitoring provided by Western journalists – mostly from major German media but also from the Kyodo News Agency (Japanese) or the *Washington Post*" (Escobar 2014c).

## War crimes

Two Russian journalists were killed by mortar shells fired from Ukrainian government positions as the Ukrainian army attacked the rebel-held town of Metalist in the Luhansk region. Eye-witness reports indicated that they had been deliberately targeted. Grey considered "their deaths were the latest in a series of anti-Russian provocations carried out by the Western-backed regime in Kiev" (Grey 2014). This occurred shortly after pro-Russian separatists in Luhansk, the easternmost region of Ukraine bordering Russia, shot down a government military transport, killing all forty soldiers and nine crew members.

The Kiev government launched a major offensive in April against the regions of the Donbass now under control of the separatists. At least 950 civilians died in this phase. At every stage, it seemed the Ukrainian government was working closely with Berlin and Washington (Dreier 2014). When a missile hit the regional administration building in early June, WMM dutifully reported the Kiev government's version of events as fact, namely that the anti-Kiev forces had blown up their own headquarters, just as several weeks earlier Kiev had made the equally outrageous claim that anti-Kiev forces in Odessa had set themselves on fire, and just as in July, WMM would favor Kiev's insistence that pro-Russian separatists, with no motivation, had shot down the civilian international flight MH17. It is clear, McAdams wrote, "the Kiev government lied about the attack on Lugansk [Luhansk] and that those lies were accepted as fact by the US government and the vast majority of the US mainstream media" (McAdams 2014). Kiev forces had also bombed hospitals, kindergartens and residential areas. None of these atrocities was condemned by Washington, which in 2003 had berated Saddam Hussein for having (purportedly) used poison gas on his own people, part of the US pretext for the invasion and occupation of Iraq. In July, HRW directly reported war crimes by both sides against civilians (HRW 2014), although its criticisms favored Kiev – which it clearly considered a legitimate power – and focused more on the actions of separatist forces (although in reporting on the fighting months later, HRW fingered Kiev as the principal culprit – see below). Washington called for more sanctions against Russia, which most EU countries, including Germany, willingly joined, even though European economies were negatively impacted by the loss of business and the continent was drifting back into recession (which it touched by January 2015).

Stockman stated the obvious:

> Kiev's marauding army and militias would come to an instant halt without access to the $35 billion of promised aid from the IMF, EU and US treasury. Obama just needs to say "stop." That's it. The civil war would quickly end, permitting the US, Russia and the warring parties of the Ukraine to hold a peace conference and work out the details of a separation agreement.
>
> (Stockman 2014)

Early in July, amid "a virtual blackout in the American media" (Lantier 2014c), and following weeks of artillery bombardment, Ukrainian regime forces retook the cities of Slavyansk and Kramatorsk, which occupied strategic positions along roads from Western Ukraine to Donetsk. They then marched on the major cities of southeastern Ukraine, Donetsk and Luhansk. The Ukrainian National Security and Defense Council spokesman, Andriy Lysenko, warned that Kiev would take no prisoners among opposition fighters and dismissed any possibility of creating a corridor that would allow Donetsk People's Republic (DPR) forces to leave the region.

The attacks forced a massive dislocation of citizens. According to the UN High Commission on Refugees (UNHCR), 110,000 people had fled to Russia and 54,000 were internally displaced as of June 27. HRW condemned the illegality of Ukrainian actions in Eastern Ukraine and blamed Kiev for the rising death toll in the Luhansk region, which by local estimates had reached 300 since May. HRW visited the city on August 20–22, interviewing locals who had been impacted by the heavy bombardment, noting that people had cowered in fear in basements for weeks without water or electricity. Many attacks failed to distinguish between civilians and combatants, as was evident in the use of weaponry not designed for areas where precision was required. HRW findings indicated that a large majority of attacks on the city were carried out by Ukrainian government forces, while the insurgents were responsible for far fewer and targeted government positions on the outskirts of Luhansk.

The seriousness of the setbacks to ethnic Russian separatists notwithstanding, the Ukrainian army victories appeared short-lived as, within days, alternative sources proclaimed Kiev's strategy to be in tatters "because a highly-motivated and adaptable militia has trounced Obama's troopers at every turn pushing the Ukrainian army to the brink of collapse, causing the Ukrainian Army to retreat almost everywhere" (Whitney 2014e). The Ukrainian reversal gave rise to rumors of a mooted anti-Poroshenko coup by outraged Nazi nationalists. By the end of October, elections in the eastern oblasts of Donetsk and Luhansk created the first parliament of the Donetsk People's Republic and its Luhansk equivalent. The republics would serve as de facto autonomous regions, and also the potential governments if they were to secede outright. Kiev's response, early November, was to order a halt to all transfers of public funds to the areas dominated by separatists in Eastern Ukraine, thus impacting on schools, hospitals, government agencies and state enterprises (Schwarz 2014b). Many of the responsibilities abandoned by Kiev were assumed by Russia. By the end of 2015, according to a report in the German tabloid *Bild*, apparently on the basis of German intelligence sources, and summarized in Ukraine's propaganda news agency UNIAN, Russia was spending approximately $73 million a month to pay public service salaries and pensions. Beneficiaries included over a million pensioners, "Russian" and Ukrainian fighters. In addition, Russia likely shouldered the expenditure of subsidized gas, fuel, oil, food, humanitarian payments and ammunition (UNIAN 2016b).

WMM coverage during 2014 and 2015 (I deal with the Minsk agreements in Chapter 8) as this concerned Crimea, Odessa and the Donbass consistently presumed – quite unreasonably – that the post-Yanukovych regime of Kiev was a normal, legitimate State. WMM underestimated or denied the fascist elements both outside of as well as incorporated within the Ukrainian armed services and whose voices contributed to the initial legislation that disenfranchised Russian language, culture and media and the country's two largest political parties. There was little or no consistent coverage of the extent to which Ukrainian politics was the plaything of a handful of powerful oligarchs. WMM did not acknowledge the strength of pro-Russian sentiment in Crimea, understated the legitimacy of the presence of Russian troops in Crimea, and unreasonably withheld credence from quite robust referenda processes in Crimea that called on the Crimean people to determine if they wished to secede from Ukraine and to seek annexation to Russia. In the face of Kiev's devastating military aggression against its own people in the Donbass, and irresponsible withdrawal of social security – when all Kiev had to do to secure peace was to negotiate a reasonable, pluralist or federalist solution – WMM could only find fault with the self-declared republics of Donetsk and Luhansk and embrace Kiev's assertions about Russian manipulation in Eastern Ukraine.

# 7 Representation, misrepresentation and the demonizing of Vladimir Putin

Illustrating the BBC's seamless fusion of journalistic commentary with propaganda, Cromwell and Edwards refer to the contention of Mark Mardell, the BBC's North America editor that:

> Putin and Obama are such different characters.
>
> One is a small, strutting hard man with a passion to re-create an old empire, the other the professorial president – concerned not to repeat the mistakes that happen when America behaves like an empire – ever-hesitant to use the massive brute force his country can muster.
>
> (Cited by Cromwell and Edwards 2014)

Would a senior BBC editor like Mardell, asked Media Lens, ever describe a US president or UK prime minister as a "strutting hard man"?

In this chapter I consider WMM demonization of Vladimir Putin as conflict propaganda and its apparent intensification in 2014 and 2015 (overshadowing earlier blips, as in the Russo-Georgian conflict of 2008). Anti-Russian propaganda did not end with the fall of the Berlin Wall. Common memes in Hollywood movies changed from staple depictions of a cold, concrete, totalitarian warren of spies and nuclear threat to others just as negative: psychopathic black marketeers dealing in dangerous weapons and drugs provided a staple replacement (Boyd-Barrett *et al.* 2011). Angela Borozna documented the increasing anti-Russian tone of US political commentary since 2000:

> Russia made significant changes from the Soviet totalitarian system, but instead of acknowledging this progress, the country is characterized by simplistic and misleading historical analogies associated with the KGB, the Soviet Union, and repressive government control. The opinions of Russian citizens on their political system or their president, as well as the actions by the Russian state that do not fit the description of "dictatorial power" are typically omitted from Western media coverage. The result of this selection bias builds up over time to make an overall negative image of the country and its president.
>
> (Borozna 2015)

A more balanced assessment of how Putin was viewed by Russian people came from Russia observer and Oxford University lecturer Karen Hewitt:

> From the point of view of ordinary Russians he (Putin) brought stability, coherent policies and a doubling, trebling, of their personal incomes and the wealth of Russia. He made the oligarchs pay taxes, he broke up some of the illegal empires of energy wealth and nationalized them. He dealt with local warlords in distant parts of the Russian federation and insisted that governors should be appointed from Moscow, so that he could keep an eye on them. With the oil money his government built up a huge emergency reserve so that it would never again be the basket-case of world economies, and he and his ministers devised and largely carried out federal-wide improvement projects in Education, Health, Agriculture and Infrastructure. People are hugely better off than they were, and although of course there are plenty of poor people, they too are better off than they were. [...] Not all of this is attributable to the policies of the Government, still less to President Putin – oil prices are important – but from the ordinary Russian's point of view, they have much to be thankful for under Putin.
>
> (Hewitt 2015)

A reorientation of Western attitudes to Russia and to Putin was set by 2008, Hewitt notes, when the West determined that Russia, after the five-day Georgian–Russian War in 2008, was at fault. There was little detailed examination of the sources of the conflict. Instead, WMM took the side of Georgian President Mikheil Saakashvili (later installed as governor of Odessa by the US-backed coup regime in Ukraine), and disregarded the contrary evidence presented by Russia to the UN. Even as Russia recovered from the 1990s economic turmoil, restructured its economic institutions to better fit the Western-led global capitalist order, and took control over the excesses of the oligarch classes whose new wealth the West had helped purloin during the 1990s privatization of State assets, the West accelerated its co-option of the old Eastern Europe and some previous countries of the old Soviet Union.

Writing in the *Daily Mail* (UK), the columnist Peter Hitchins lambasted his colleagues for burying the visit of the director of the CIA, John Brennan, to Kiev in mid-April, a clear indication of Washington's involvement and the likelihood of covert operations, whether conducted directly by the Washington intelligence apparatus or its private contractors. WMM coverage invited the view that US intelligence had played no role in the *coup d'état*. Blunden cited the former CIA Officer John Stockwell as saying that: "stirring up deadly ethnic and racial strife has been a standard technique used by the CIA" (Blunden 2014). Referring to media protestations, based on fabricated evidence, against the possible presence of Russian spies in the Ukraine, Lantier complained that:

> CIA Director John Brennan went to Kiev a week ago, though he sought to hide his visit from the public, as the Western-backed regime in Kiev

> prepared its crackdown on the eastern Ukraine protests. British intelligence has admitted that its agents are combing east Ukraine. Why is the dispatching of spies to Ukraine by Russia more threatening than the appearance of MI6 or of Brennan, who has played a leading role in running a global network of torture camps and a program of drone murder?
>
> (Lantier 2014a)

WMM often referred to those who resisted rule by the coup regime from February 2014 as "Ukrainian pro-Russian separatists." They might more accurately have been described as "federalists of southeast Ukraine" (McGovern 2014a), although the term "ethnic Russians" could also suffice to denote the community from which many of the separatists came. Parry described an "odor of prejudice in how the mainstream US news media treats the ethnic Russians in eastern Ukraine, as if they are mindless beings, easily duped 'minions' of Vladimir Putin" (Parry 2014i). A *New York Times* editorial of May 12 diminished the Eastern Ukrainian rebel leaders as "self-declared," denied there was backing for the rejection of Kiev-appointed regional leaders, and mocked claims of legitimacy for the separatist referendums in Donetsk and Luhansk. The *New York Times* editors bloviated that: "If there were questions about the legitimacy of the separatist referendums in eastern Ukraine, the farcical names of the entities on which people were asked to vote – the self-declared People's Republics of Donetsk or Luhansk – surely answered them" (*New York Times* Editorial Board 2014a). In face of contrary evidence, including a *New York Times* reporter's account of Ukrainian National Guard attacks on voting booths – the paper attributed all violence to the rebels. Violence that was the result of pro-Kiev New National Guard militants was attributed instead to the ethnic Russian separatists. The *New York Times* persistently saw Putin's hand behind the separatists, a scenario that even the veteran policy adviser Henry Kissinger declared impossible. Referenda in the provinces of Donetsk and Luhansk obliged a *New York Times* reporter to acknowledge "substantial popular support for the pro-Russian separatists in some areas" (Kramer 2014a).

## Creating "facts"

Politicians and the press struggled to transform the ambiguities and falsehoods into "facts" that they then embedded into the press archives to be resurrected as "givens." On September 1, the German President Joachim Gauck accused Russia of conducting an "offensive struggle for new borders and a new order" based on the "law of the jungle." With breathtaking selectivity, given German passivity through the US invasions of Afghanistan and Iraq, and its participation through NATO in the destabilizations of Libya and Syria, Gauck asserted that Germany would "resist those who breach international law, annex foreign territory, and militarily support divisions in foreign lands" (Schwarz 2014a). In the US press, journalists increasingly referred with little or no qualification to what they called Russian aggression against the Ukraine, the Russian "seizure" of Crimea; and pro-Russian separatist "responsibility" for the shooting down of MH17.

Faithful to the established propaganda practice of channeling hate by means of personalization, simplification and iconicization, WMM's focus on a single leader dominated their narrative, portraying Putin as an imperialist with aggressive designs on Ukraine, even as a dictator (despite overwhelming contrary evidence). This process of demonization had come into vogue in 1980s Washington in order to promote US armed interventions in Central America:

> Since nearly all foreign leaders have serious flaws, it proved much easier to demonize them – and work the American people into war frenzies – than it was to persuade the public that Washington's favored foreign leaders were actually paragons of virtue. Once the black hat was jammed on a foreign leader's head, you could say whatever you wanted about him and disparage any American who questioned the extreme depiction as a "fill-in-the-blank apologist" or a "stooge" or some other ugly identifier that would either silence the dissenter or place him or her outside the bounds of acceptable debate.
>
> (Parry 2015i)

Parry worried that:

> The demonization of Putin in the Western media has been so total that anyone who dares question the most extreme interpretations of his behavior is denounced as a "Putin apologist." Indeed, any attempt to present a nuanced narrative of what has happened in Ukraine is dismissed as somehow promoting Russian imperialism or spreading Russian propaganda. This oppressive "group think" has, in turn, made formulating any rational policy toward Russia and Ukraine politically impossible in Official Washington.
>
> (Parry 2014g)

WMM repetitively described Putin as authoritarian, "despite the fact that he has been repeatedly elected by large majorities in competitive elections against Western backed and funded candidates" (Petras 2015). His popularity in Russia was attested to polls conducted by Western agencies. Public support for Putin soared to 83 percent in 2014 and over 85 percent in 2015 (Ray and Esipova 2014). While Western-imposed sanctions and a decline in oil prices created severe problems for the Russian economy in 2015 onwards, the economy and the ordinary people of Russia continued to be a great deal healthier and safer than in the lawless days under perpetually boozy, Western-supported Boris Yeltsin in the 1990s.

WMM persistently described Putin and Russia as aggressive, yet Russia's military budget remained at one-tenth that of the USA, supporting a far smaller number of bases at home or abroad, that were typically much smaller than US bases (Margolis 2015). Certainly, Putin protected Crimea which had voted to secede from the Ukraine and join Russia. Russia's security interests in Crimea, and outstanding treaty rights to the maintenance of Russian troops in Crimea, were never seriously in doubt. Yet Putin was cautious not to inflame the hopes of ethnic Russians in Eastern Ukraine opposed to the coup regime, some of whom desired annexation with Russia. Putin was conciliatory, offering and actually moving Russian troops back from the Russian border and requesting the upper house of the Russian federation to withdraw its authorization to use force to protect Russian security. Following the election of Poroshenko as Ukrainian president, the *Wall Street Journal* indicated that Putin would stand down from supporting resistance in Eastern Ukraine (Troianovski and Lee 2014). Moscow returned its ambassador, Mikhail Zurabov, to Kiev for inaugural festivities, and Russian news agencies reported that Putin had tightened border controls to prevent people from crossing to fight against Ukrainian forces.

A speculative *New York Times* piece in April (Baker 2014) discussed allegations of Putin's "secret fortune," reminiscent of how WMM had castigated the supposedly lavish living of ousted Ukrainian President Viktor Yanukovych. WMM's obsession with the wealth of regime leaders that Western states seek to discredit is the hallmark of official hypocrisy, coming as such accusations do from the very sources of neo-conservatism that precipitated the world towards the greatest wealth inequalities that the world has ever seen – so that by 2014 only eighty-five people controlled half of the world's wealth (Wearden 2014), a number that declined to sixty-one in 2016.

WMM less frequently attacked Putin where perhaps he was most vulnerable, namely his insistence on Russian media "patriotism" in their coverage of the wars in Chechnya in the 1990s. The Chechen wars, on the other hand, were something of a Pandora's Box that might raise questions about Western intelligence maneuvers to channel mujahidin from Afghanistan and the Middle East into the former Soviet Union for the purposes of destabilization (Ahmed 2006). Putin's signing of a law in December 2013 that would make the spreading of separatist views a criminal offence, punishable by up to five years in jail, also passed underreported.

## Ignoring Putin the peacemaker

WMM gave scant coverage to Putin's speech to the Organization for Security and Co-operation in Europe (OSCE) on May 7, which proposed that the only option was "to launch direct dialogue, genuine, full-fledged dialogue between the Kiev authorities and representatives of southeast Ukraine. This dialogue could give people from southeast Ukraine the chance to see that their lawful rights in Ukraine really will be guaranteed" (Putin 2014). He offered to withdraw Russian forces from the border with Ukraine (and troops were pulled back). Putin also asked representatives of the southeastern regions of Ukraine to postpone the referendum scheduled for May 11 (it proceeded anyway).

WMM frames did not include Putin-as-peace-maker (Whitney 2014b), irreconcilable with their preferred demon frame or, more importantly, Washington policy. Washington wanted war, not peace:

> War is the vehicle for breaking up the Russian Federation into tiny statelets that pose no threat to US military bases spread throughout Asia. War is the means by which Washington can make its pivot, surround China, and control its future growth. War paves the way for establishing US outposts in Ukraine and subverting greater economic integration between Russia and Europe.
>
> (Whitney 2014b)

Despairing of WMM, the veteran US investigative reporter Robert Parry wrote:

> Sometimes dealing with the waves of US media propaganda on the Ukraine crisis feels like the proverbial Dutch boy putting his fingers in the dike. The flood of deeply prejudiced anti-Russian "group think" extends across the entire media waterfront – from left to right – and it often seems hopeless correcting each individual falsehood.
>
> The problem is made worse by the fact that the NYT, the traditional newspaper of record, has stood out as one of the most egregious offenders of the principles of journalism. Repeatedly, the Times has run anti-Russian stories that lack evidence or are just flat wrong.
>
> (Parry 2014d)

Anti-Russian sentiment in the *New York Times* hardened over time. After the February coup, the *New York Times* reported that among the eighty people killed in Maidan were more than a dozen police officers. Only later did the storyline change to: "More than 80 protesters were shot to death by the police as an uprising spiraled out of control in mid-February" (*New York Times* March 5). Both dead police and the contested circumstances surrounding the sniper fire simply disappeared from the *New York Times*' narrative:

> It became flat fact: evil "pro-Yanukovych" police gunned down innocent "pro-democracy" demonstrators. Also consigned to the memory hole was the key role played by well-organized neo-Nazi militias that led the final assaults on the police.
>
> (Parry 2014d)

When the *New York Times* wanted to discredit Putin, it told readers he had lied when he claimed there were no Russian troops in Crimea and then later acknowledged their existence. Yet, as Parry points out, the Russians never denied having troops in Crimea, where they maintained the Black Sea naval base in Sebastopol and had a contract with Ukraine allowing up to 25,000 troops. Elsewhere, Parry recorded "new levels of extreme," functioning as conduits for neocon propaganda at the US State Department:

> Everything that the Times writes about Ukraine is so polluted with propaganda that it requires a very strong filter, along with additives from more

> independent news sources, to get anything approaching an accurate understanding of events.
>
> (Parry 2014j)

Howard Friel recalled "that from September 2001 to March 2003, a period in which the Bush administration repeatedly threatened to invade Iraq, the *New York Times* never mentioned the words 'UN Charter' or 'international law' in any of its seventy editorials on Iraq" (Friel 2014). By contrast, it stirred up an indignant froth about Russia's "invasion" of Crimea – where no civilians were killed, nothing was destroyed, and where a large majority was soon to vote for incorporation into Russia. WMM painted the crisis as Russian "imperialism." In real life, Moscow reacted cautiously to aggressive US promotion of NATO expansion to Russia's border in violation of its post-Cold War pledge (McGovern 2014a). Russia's acceptance of a request from the Crimean people for annexation with Russia was less aggressive than NATO's support for the cessation of Kosovo from Serbia:

> Our western partners created the 'Kosovo precedent' with their own hands. In a situation absolutely the same as the one in Crimea they recognized Kosovo's secession from Serbia as legitimate while arguing that no permission from a country's central authority for a unilateral declaration of independence is necessary. [...] And the UN International Court of Justice agreed with those arguments. That's what they said; that's what they trumpeted all over the world and coerced everyone to accept – and now they are complaining about Crimea. Why is that?
>
> (Whitney 2014f)

WMM was determined to find evidence of Russian agitation behind the unrest. Notwithstanding the paper's retraction of photos it falsely alleged had proven the presence of Russian troops, the *New York Times* was back by early May, undeterred by its own finding that fighters of the People's Militia in Donetsk were indeed Ukrainian, and veterans of – but not active duty members of – the Russian armed forces. The paper insinuated that:

> motley assortments of fighters from other war zones that are intimately associated with Russia would be unlikely to surface against the powerful will of the Russian president, Vladimir V. Putin. [..] [raising] the possibility of a more subtle Russian game that could keep Ukraine unbalanced for years.
>
> (*New York Times* cited by Hart 2014)

Peter Hart recaps:

> There is no evidence that Russia is in control of any of this, but the lack of such evidence may be a sign of a "more subtle" game. [...] What you're left with from the Times is the suggestion that the lack of direct evidence

> is probably proof that Russia is up to something – i.e., "leaving no fingerprints."
>
> (Hart 2014)

Anything Putin had to say, even when incontrovertibly accurate, was deemed by WMM to constitute propaganda. Anything Washington had to say was taken at face value. Parry begged:

> look what happens when Russia's President Vladimir Putin does what the U.S. news media should do, i.e. point out that "It's necessary to recall the actions of the United States in Afghanistan, in Iraq, in Libya, where they acted either without any sanction from the U.N. Security Council or distorted the content of these resolutions, as it happened in Libya. There, as you know, only the right to create a no-fly zone for government aircraft was authorized, and it all ended in the bombing and participation of special forces in group operations." Despite the undeniable accuracy of Putin's observation, he was promptly deemed to have "lost touch with reality," according to a Washington Post's editorial, which called his press conference "rambling" and a "bizarre performance" in which his words have "become indistinguishable from the propaganda of his state television network."
>
> (Parry 2014a)

Events that would cast any Western leader as strong and wise were routinely misinterpreted in the case of Putin. His bold decision to send Russian bombers to support Assad in Syria (and, by implication, help stabilize Iraq and support Iran) against ISIS and ISIS oil routes to Turkey and, through Turkey, to Europe quickly helped expose the secret, corrupt complicity of Western powers and their reactionary allies in Saudi Arabia, Qatar and the UAE, with so-called "moderate" Syrian jihadists and with ISIS itself. WMM blanketed Russia's intervention with reports of sudden though questionable US "successes" or scrupled over alleged Russia-attributed civilian casualties that would cause them little concern if attributable to Western powers. In covering civilian casualties of US military actions (as in the case of the bombing October 3, 2015 of a Médecins Sans Frontières [Doctors without Borders] hospital in the Afghan city of Kunduz), WMM uncritically deferred to Pentagon sighs over regrettable "errors." A 2015 Russian documentary, *World Order*, on the Russian government thinking about the Middle East was largely ignored or misunderstood in WMM (Doctorow 2016). That Russia had correctly identified NATO as a security threat to Russia, was highlighted, and even exaggerated. Underplayed or ignored was Putin's insistence that: Russia did not want the Soviet Union back; nuclear war was necessarily catastrophic; and he could not imagine anyone using it. The documentary critiqued the US struggle for global hegemony on the basis of pretexts such as democracy promotion and humanitarian intervention, amidst preposterous insistence on American "exceptionalism." It examined the chaotic aftermath of "color revolutions," and the impunity of Western leaders for their war crimes. Several former Western leaders even voiced support in the documentary for such charges (Doctorow 2016).

## Accelerating the propaganda

From early in 2015, WMM framing of Russia as an aggressor and Putin as a crafty dictator grew more virulent and less connected to known facts. The tone was set by November's Australian G-20 Summit in the light of Donetsk's effective secession from Ukraine. All parties issued public statements indicating their review and augmentation of respective capacities for conventional and nuclear war. Russia was threatened; Putin was verbally abused by other world leaders and treated like a leper:

> Australian Premier Tony Abbott, declared that he would "shirtfront" Putin. At the meeting, Canadian Prime Minister Stephen Harper reluctantly shook Putin's hand while telling him, "You have to disappear from Ukraine." [...] US President Barack Obama accused Putin of violating international law. "You do not march into other countries," Obama declared [...] Ukrainian President Petro Poroshenko threatened Russia with total war.
>
> (Schwarz 2014b)

James Bissett, the former Canadian ambassador to Yugoslavia, Albania and Bulgaria wrote in the *Ottawa Citizen* that:

> The current crisis in Ukraine threatens global security and at worst has the potential for nuclear catastrophe. At best, it signals a continuation of the Cold War. Sadly, the crisis is completely unnecessary and the responsibility lies entirely in the hands of the United States-led NATO powers. The almost virulent propaganda onslaught blaming Russia for the instability and violence in Ukraine simply ignores reality and the fact.
>
> (Bissett, cited by Deutsch 2015)

In its first Sunday edition of 2015 the *New York Times* ran a front-page story attributed to Andrew Higgins and Andrew Kramer headlined "Ukraine Leader Was Defeated Even Before He Was Ousted," arguing that the events of Kiev in 2014 amounted to a "meltdown," not a coup (Higgins and Kramer 2015). It was based on interviews with former police officials and others who had abandoned Yanukovych who – according to the article – was "cast adrift" by his allies. Higgins had previously co-authored a *New York Times* story with fabled Michael Gordon (co-author with Judith Miller of many of the pre-2003 *New York Times* stories that helped pave the way to the invasion and occupation of Iraq) that used faked photo evidence in support of claims that the Russians were responsible for the uprising in Eastern Ukraine. The story effectively whitewashed US involvement from the events (Van Auken 2015a). It made no mention of State

Department's Victoria Nuland, whose leaked telephone call with Geoffrey Pyatt, US ambassador in Kiev, showed her scheming who would be in power following the toppling of Yanukovych; nor mention of the $5 billion that Nuland had informed the world, the USA had spent on the faux-NGO National Endowment of Democracy (among others) to meddle with Ukraine's internal politics. The backgrounds of many key informants were obscured: Sergei Pashinsky (caught with a sniper's rifle in Maidan) and Andriy Parubiy (fanatic supporter of the fascist, Stepan Bandera, and founder of the neo-Nazi party, Svoboda, which was rewarded after the coup with three ministries and three governorships). Not until the following September were US audiences, via CBS's *60 Minutes*, entrusted with the viewpoint of Russian President Vladimir Putin:

> Putin added that he "knows for sure" that the US had in some way helped oust Viktor Yanukovych. Moscow knows "who had meetings and worked with people who overthrew" the ex-president, as well as "when and where they did it," he said.
>
> "We know the ways the assistance was provided, we know how much they paid them, we know which territories and countries hosted trainings and how it was done, we know who the instructors were. We know everything. Well, actually, our US partners are not keeping it a secret," he said.
>
> "Since the start of the Ukrainian conflict the West has cited a "Russian threat" as an excuse to justify increased military spending in Eastern European NATO countries and provide a reason for US troops to be stationed in Europe. The presence of Russian troops on Russia's territory near the border with Ukraine has been used to argue that Russian troops are involved in the Ukrainian conflict."
>
> (RT 2015h)

Bloomberg's Josh Rogin in a report headlined "Inside Obama's Secret Outreach to Russia," asserted that the Russian President Vladimir Putin showed little interest in "halting his aggression in neighboring Ukraine" (cited in Zuesse 2015a). Zuesse noticed that Rogin drew (almost) solely from sources inside the US Administration that could easily have been refuted with a wider range. As tensions between the USA, EU, Russia and Ukraine escalated into February, European and US media ratcheted up anti-Russian hostility, working relentlessly to "pollute public opinion" (Lantier 2015b). Western "liberal" media played similar games as their conservative peers: *Le Monde* blamed the Ukraine crisis squarely on Vladimir Putin. Robert Parry abhorred the excruciating condescension of WMM, citing the title of an article by the *Washington Post*'s Fred Hiatt (early advocate of the US invasion of Iraq in 2003 on the pretext of WMD): "Mr. Putin's revisionism: His paranoia shouldn't blot out the good the West tried to offer," and online as "After the fall of the Soviet Union, the US tried to help Russians" (Parry 2015c).

PBS aired a one-hour exposé Putin's Way on *Frontline* in January 2015. Created by the writer, director and producer Neil Docherty, it featured the work of the investigative reporter Gillian Findlay. It castigated Putin for assisting in and profiting from the process of converting State assets from the Soviet Union to private assets during the regimes of Gorbachev and Yeltsin. Yet these policies were wholeheartedly promoted by the USA and EU as "shock therapy": the process was conceived and orchestrated in Washington.

> When a conflict erupted in 1993 between the Yeltsin regime and the parliament over these deeply unpopular policies, the Russian president bombarded the parliament building, killing hundreds. US President Bill Clinton endorsed these actions, declaring, "President Yeltsin had no other alternative but to try to restore order." At that time, Washington had no objection to the profoundly anti-democratic methods of the new Russian ruling elite.
>
> (Peters 2015b)

Ostensibly concerned for democracy in Russia, the documentary worried about the role of oligarchs: not oligarchs in general, only those who were not favorably disposed towards Washington. Mikhail Khodorkovsky appeared in a positive light. Khodorkovsky, once Russia's richest man worth $15 billion, was tried and jailed for ten years by the State for fraud, embezzlement and money laundering. The US-German author William Engdahl considered Khodorkovsky's real offences were that he was "a key part of a Western intelligence operation to dismantle and destroy what remains of Russia as a functioning state" (Engdahl 2011) and was in the process of buying enough seats in the 2004 Duma elections to enable him to change Russian laws regarding the ownership of oil and pipelines. He was negotiating the sale of a large part of Yukos to one or more Western oil companies.

The documentary's controversial narrative of Putin's access to the presidency (i.e., turning terrorist attacks in Moscow into a pretext for a second Chechen war) made no reference to possible false flag involvement of Russian security services or to the comparable controversy surrounding the access of George W. Bush to the presidency and his exploitation of the 9/11 terrorist attacks to pursue an imperialistic foreign policy. Nor was reference made to neo-conservative involvement in the *coup d'état* of EuroMaidan in 2014. Peters concluded that: "Putin's Way is not so much a documentary as a propaganda piece intended to justify Washington's confrontation with Russia and make the case for regime-change in Moscow, if necessary by military means" (Peters 2015b).

So hysterical had the propaganda environment become that on the eve of the Minsk II talks in February, the co-editor of conservative German newspaper *Frankfurter Allgemeine Zeitung* (FAZ) called on NATO to militarily escalate the confrontation to the point of nuclear attack (Schwarz 2015). Upon completion of the Minsk talks on February 11 (leading to renewal of the first Minsk agreement), Andrew Higgins of the *New York Times* declared that Europe's "often fractious" leaders had "confronted the reality that Mr. Putin retains the upper hand precisely because he is prepared to use military force to get what he wants in diplomacy" (Higgins 2015). Higgins neglected to mention that Europe's leaders were "fractious" because they were under US pressure, at high actual and potential cost – even possible nuclear conflict – to maintain an aggressive posture towards Russia. That the USA threatened to arm Kiev – a move calculated to accentuate the risks – was whitewashed out. Yet the only reason why Putin declared himself ready to use military force – the likelihood of an actual Russian presence in Eastern Ukraine was still small, despite NATO's vehement protestations to the contrary – was because the USA was threatening to arm Kiev.

Higgins' story served as US State Department propaganda. It scapegoated Putin as the one person mainly responsible for the breakdown of the previous September's truce whereas blame could better be attributed to Kiev's failed attacks on the new republics of Luhansk and Donetsk and its withdrawal of all aid from citizens of territories controlled by separatists. The breakdown of Minsk I had cost Ukraine an additional 200 square miles ceded to the separatists. Higgins sought further to discredit Putin, citing a US intelligence officer who accused him of playing "a long game," a US projection that attributed to Putin strategies that better characterized Washington neo-conservative maneuvers. Without proof, the *New York Times* asserted that Russia had been fortifying ethnic Russians with weapons. It cited a critical source who whined that Putin was not "showing his cards" (whereas the West, of course, always does!). The Higgins story found Putin further at fault because his objectives in Ukraine were "vague" other than that he wanted to block NATO from a perceived plot to penetrate further into former Soviet territory. This plot was quite easy to substantiate, since Zbigniew Brzezinski and leading US neo-conservatives had talked about it over two decades. The mischievous Putin had apparently "made clear" that he wanted to divide the EU, even though the latter had remained "unusually united." Not a single pro-Russian source nor even a single Russian source was cited.

Particularly among more popular WMM, anti-Putin coverage verged on the silly, with allegations, based on next to no evidence nor even a minimally sensible rationale, that Putin had hijacked the missing MH370 airline and landed it in Kazakhstan (covered in *New York Magazine*, the *Daily Mail* and CNN, among others) or that, according to reports of a "Pentagon think-tank," Putin was autistic (Malic 2015). A front-page *New York Times* story in June by Peter Baker and Steven Erlanger was titled "Russia Uses Money and Ideology to Fight Western Sanctions." It warned "Moscow has brought to bear different kinds of weapons, according to American and European officials: money, ideology and disinformation." This attracted the derision of Robert Parry who, after chastising what he considered its lack of hard evidence, marveled:

> how it ignores the much more substantial evidence of the US government and its allies themselves financing propaganda operations and supporting "non-governmental organizations" that promote the favored US policies in countries around the world. [...] Plus, there's the failure to recognize that many of Official Washington's own accounts of global problems have been riddled with propaganda and outright disinformation.
>
> (Parry 2015f)

Parry called attention to the hundreds of millions of dollars disbursed by the USA to finance "journalism" organizations, political activists and "non-governmental organizations" that promoted US policy goals inside targeted countries including scores of such operations in Ukraine financed by the NED, whose budget from Congress exceeded $100 million a year (Parry 2015j).

## The Nemtsov assassination

Demonization took a more sinister turn following the assassination of Boris Nemtsov, leader of the Russian opposition, on February 28, 2015, prior to a planned demonstration in Moscow in protest against Russia's Ukraine policies. Nemtsov had been a deputy prime minister under Boris Yeltsin and was favorably regarded by the former British Prime Minister Margaret Thatcher. Lendman (2015) described him as a RPR-Parnas party co-chair (a party with 5 percent of the popular vote against Putin's 85 percent), a Yaroslavi Oblast regional parliament member, and Solidarnost co-founder/co-chair – modeled after CIA-financed anti-communist Lech Walesa's Polish Solidarnosc. Before the 2014 coup in Kiev, Nemtsov supported Ukraine's pro-Europe orientation. He, too, was a beneficiary of funding from the National Endowment for Democracy (NED). With Garry Kasparov, he had met with US President Obama in 2009. Lendman speculated that Nemtsov's assassination was a false-flag killing since Putin had nothing to gain, whereas Western powers stood to gain a great deal. Roberts (2015) allowed that the killing might have been staged by Russian nationalists. Charges by the Russian investigation committee were brought in January 2016 against a number of individuals including Ruslan Mukhutdinov. Motivations were unclear. Rumors that the assassination was a reprisal for Nemtsov's alleged anti-Islamic statements were not widely supported at the time.

The killing may have constituted a form of propaganda-by-deed, and likely provided timely credibility of Western claims for the benefit of their own publics of the Russian administration's proclivity to aggression. WMM pointed the finger of culpability towards Putin citing previous (but somewhat dated) murders frequently attributed (with greatly varying evidence) to state agencies. These included: the liberal politician Sergei Yushenkov (2003); the investigative journalist Yuri Shchekochikhin (2003); *Forbes* magazine's Russian editor Paul Klebnikov (2004); the investigative journalist Anna Politkovskaya (2006); the former Russian spy Alexander Litvinenko (2006); the human rights campaigner Natalia Estemirova (2009); and Boris Berezovsky, the former Kremlin power broker turned Putin critic (2013). Such coverage quoted known anti-Putin figures such as Mikhail Khodorkovsky, one-time gas oligarch (jailed for ten years and now in exile) and even the former world chess champion Garry Kasparov. Kasparov, a right-wing anti-Putin member of the US foreign policy establishment, wrote an opinion column in the following Monday's *Wall Street Journal* in which he called for the US to treat the Putin administration as a "criminal rogue regime." Oft-quoted by WMM were current opposition leaders such as Ilya Yashin, Mikhail Kasyanov, Alexei Navalny and Gennady Gudkov. Many of these figures had been funded by the US State Department-backed NED or by an NGO funded by NED.

Citing a different possible motivation, Ukraine's Petro Poroshenko claimed that Nemtsov was about to reveal evidence of direct Russian support for the uprising in Eastern Ukraine. For others, Nemtsov was targeted either for his symbolic importance or for his useful dispensability. Rivals such as Alexei Navalny may have been considered more suited to a leading role in any NED-orchestrated Russian "Maidan." Navalny, described by Cartalucci (2014b) as a neo-fascist, co-founded the Democratic Alternative movement (DA!) and was vice-chairman of the Moscow branch of the political party Yabloko. DA!'s leader, Mariya Gaydar also received funding from the NED and worked with Ilya Yashin of NED-funded People's Freedom Party (merged with RPR-Parnas in 2012). Yashin was close to Vladimir Ryzhkov, a founding member of RPR-Parnas and a member of NED-funded Washington-based World Movement for Democracy. He was a leading member of the Strategy 31 campaign for free assembly whose ranks, claimed Cartalucci, "are filled with activists trained and coordinated by US NED-funded NGOs" (Cartalucci 2014b) and whose activities had spread from Moscow, to St. Petersburg, Archangelsk, Vladivostok, Yekaterinburg, Kemerovo and Irkutsk (about twenty Russian cities in total). The NED was deleting evidence of many such connections from public documents. Cartalucci's December 2014 assessment (2014b) suggested a network of actual or potentially rebellious organizations, funded in part by the US State Department along the lines of previous color revolutions, were readying for a Moscow coup. Knowledge of this would have provoked a defensive climate among State security agencies, making some kind of pre-emptive or retaliatory action more likely.

Mikhail Gorbachev also thought the killing to be a provocation of some kind to complicate or destabilize the regime. David North, for *World Socialist Web Site* wondered if the USA was planning a palace coup that would replace Putin with a Western-friendly oligarch. North detected two potential scenarios: Nemtsov was killed by elements within the Putin faction as a reckless warning to the opposition; or, he was killed by members of the anti-Putin faction who considered Nemtsov excessively cautious and an impediment to violent regime change. They may have concluded that Nemtsov would serve the opposition better as a martyr (North 2015).

A year later Moscow authorities released the conclusion of an official investigation into his killing, finding those responsible to be Anzor Gubashev and Zaur Dadaev, both originating from Northern Caucasus and whose motives may have included a desire for retaliation for Nemtsov's allegedly anti-Islamic views. This conclusion was roundly downplayed or ignored by WMM, presumably because it did not support their preference for an outcome that would confirm Putin's satanic proclivities.

While Western media cast suspicion on President Putin specifically or on the Moscow administration more generally, they were virtually silent about a string of mysterious opposition deaths in Ukraine. These included the apparent suicide, on the same day as Nemtsov's murder, of Mykhailo Chechetov, a key ally of deposed Ukrainian president Victor Yanukovych. In separate articles William Blum, Justin Raimondo and Raul Meijer listed the following killings outside pro-Russian areas during the same time period (Blum 2015; Raimondo 2015a; Meijer 2015). Additionally, both the *Financial Times* and *France-24* used Kiev sources to imply, without evidence, that Russians were to blame for these killings in a black campaign intended to further destabilize Ukraine (Cunningham 2015b). The killings included:

*January 26*: Nikolai Sergienko, former deputy chief of Ukrainian Railways and a supporter of Viktor Yanukovych's Party of Regions, reportedly shot himself with a hunting rifle. The windows were all locked from inside, and no note was found.

*January 29*: Former Chairman of the local government of the Kharkov region, Alexey Kolesnik, hanged himself.

*February 24*: Stanislav Melnik, a member of the opposition party (Partia Regionov), shot himself.

*February 25*: The Mayor of Melitopol, Sergey Valter, hanged himself a few hours before his trial.

*February 26*: Alexander Bordiuga, deputy director of the Melitopol police, was found dead in his garage.

*February 26*: Alexander Peklushenko, former member of the Ukrainian parliament, and former mayor of Zaporizhi, was found shot to death.

*February 28*: Mykhailo Chechetov, former Member of Parliament, member of the opposition party (Partia Regionov), "fell" from the window of his seventeenth-floor apartment in Kiev.

*March 14*: The 32-year-old prosecutor in Odessa, Sergey Melnichuk, "fell" to his death from the ninth floor.

*March 22*: Yanukovych's 33-year-old son Viktor Jr., a former Ukraine MP, died after his car "apparently fell through ice on Russia's Lake Baikal."

*April 15*: Oleg Kalashnikov, yet another prominent Party of Regions leader, died of a gunshot wound.

*April 16*: Oles Buzina, a historian and journalist, was shot dead.

*April 16*: Serhiy Sukhobok, a journalist, was shot dead.

*April 17*: Olga Moroz, editor-in-chief of the *Neteshinskiy Vestnik*, found dead in her home. Her body showed "signs of violent death."

*Russia Insider* (2015) mentioned *Vesti* journalists Vyacheslav Veremii and Ruslan Kotsaba. Some murders followed the posting of personal details of victims on the site 'Peacemaker' (psb4ukr.org), a site that was supported by the

Ukraine MP and government advisor Anton Gerashchenko. Consideration was also due (Meijer 2015) to disappearances, including Dr. Skorokhodov Vitali and "militiaman" Alexey Astanin in March. In June 2015, two persons charged with the murder of Oles Buzina (see list above) were said to be neo-Nazi paramilitary militants. One of them, Denys Polischuk, was released on bail on June 23 – the bail was paid by Oleksiy Tamrazov, owner of Media Group conglomerate and a major oil/gas tycoon.

Controversies over Nemtsov spilled into allegations and counter-allegations about respective degrees of press freedom between Russia and Ukraine, and the riskiness of journalism. Ahead of the planned elections in the Donbass in 2015, Ukraine banned thirty-four foreign journalists and seven bloggers from entering the country. They were part of a larger blacklist comprising 388 individuals and over 100 organizations forbidden from entry on the grounds of "national security." The list included businesspeople and journalists from the USA. President Poroshenko averred that the move to censor international media was taken "in coordination with our partners from the European Union, the United States of America and other countries" (Raimondo 2015c). A Ukraine-friendly survey by Lyachynska (2016) found that press freedoms had improved under Poroshenko. Ukrainian journalists were doing a far better job than state institutions – namely prosecutors, judges and police – in exposing corruption. Some of Ukraine's best journalists were now in parliament. News organizations such as Slidstvo.Info, Nashi Groshi, *Ukrainska Pravda*, Radio Free Europe/Radio Liberty's "Schemes" program routinely exposed wrongdoing and even crimes. Nonetheless, six oligarchs controlled most major media outlets. They were Ihor Kolomoiski, Dmytro Firtash/Serhiy Lyovochkin, Rinat Akhmetov, Victor Pinchuk, President Petro Poroshenko and Serhiy Kurchenko (Yanukovych's alleged front man). Few respected the principle of editorial independence, less than ever in advance of elections.

On the Russian side, WMM allegations of the dangers of doing journalism in Russia and of official assassinations of journalists were exaggerated. Karlin noted that Russia had twice the number of newspaper journalists as the USA despite having half the population and that "when adjusted for per capita rates, Russian journalists have always been far safer than any number of 'democratic' countries friendly to the USA such as Brazil, Mexico, India, Colombia" (Karlin 2015). He cited a 2008 finding by blogger Fedia Kriukov that over 50 percent of the killings of Russian journalists that the Committee to Protect Journalists (CPJ) attributed to professional activities were wholly or partially falsified. Many were killed for reasons that had nothing to do with the Administration.

Russian broadcast media in 2015 were mainly dominated by two government-controlled channels, Channel One and Russia One, and a third, NTV, was owned by the State-controlled energy giant Gazprom. A leading source of satellite television news was government-controlled RT. There were two State-owned news agencies, TASS and RIA-Novosti. The most successful news agency, however, was Interfax; it is privately owned. A BBC site claimed that since the Ukraine crisis:

> Russian state media have intensified the pro-Kremlin and nationalistic tone of their broadcasts, pumping out a regular diet of adulation for Mr Putin, nationalistic pathos, fierce rejection of Western influence and attacks on the Kremlin's enemies.
>
> (BBC 2015)

The BBC site noted hundreds of radio stations and 400 hundred daily newspapers ("catering for most tastes") the most popular of which supported Kremlin policy. Several influential dailies had been bought by sources close to the Kremlin. Of eleven daily and weekly papers listed, only one was government-owned. Another was identified as left leaning, and at least two (I would propose) regularly carry articles critical of the Kremlin (*Novaya Gazeta* and the *Moscow Times*).

From the end of 2014, alternative Western media sites in the West were increasingly obliged to handle attacks on their enterprises and reputations from strategic communication (Stratcom) or Information Operations (IO) in Ukraine or the West. The Association of Old Crows' Information Operations Institute Director, Joel Harding, handled information operations in Ukraine and by February 2015 assembled an "i-army" of 40,000 to monitor the "criminals who destroyed the Ukrainian people," including journalists critical of the Kiev government. Their targets included RT, described as an alleged "propaganda" agency by leading US and UK politicians and intelligence personnel (Eliason 2015b). The libertarian news site Anti-War.com was undermined twice in 2015 by Google's Adsense which withdrew the site's advertising when Anti-War refused to remove content both on Iraq and Ukraine of which Adsense disapproved. Such efforts more than matched equivalent Russian operations in Moscow's Internet Security Agency (described elsewhere). In fall 2015, Ukrainian propaganda sources reported false (but rarely visited) Ukrainian and US websites set up by Russia to spread disinformation about Ukraine (UNIAN 2015g).

## Litvinenko

Vladimir Putin was again targeted as a murder villain by a British inquiry in January 2016. This concerned the 2006 poisoning of Alexander Litvinenko, an ex-FSB agent in the pay of British MI6. Litvinenko was a close associate of the Putin critic and oligarch Boris Berezovsky (killed in 2013) and had written books that were antagonistic to Putin and Russia.

The British report was substantially flawed. The former lawyer Alexander Mercouris and Litvinenko-expert William Dunkerley argued that inquiry Chair, Sir Robert Owen, confused his role with that of a criminal judge, pronouncing not only the cause of death but also accusing two alleged perpetrators (Russians Andrei Lugovoi and Dmitry Kovtun) whom he alleged had connections with Russia intelligence agency FSB, even claiming that the murder had "probably" been ordered by Vladimir Putin (of which there was no factual evidence). The process allowed no cross-examination and no jury. In a trial court, it would not have been possible to convict on the basis of sometimes secret evidence from anonymous witnesses, where the defendants had no right of appeal. A key witness refused to attend and the defendants were not present. They could not be extradited from Russia because extradition was contrary to the Russian constitution. Thus the evidence that convicted them went unchallenged. The Chair sometimes reached outside of the courtroom for evidence that suited his prejudices (based, his critics claimed, on a partisan, incorrect and dated understanding of modern Russia) and his evidentiary standards varied between witnesses.

The report suggested Russian non-cooperation with British authorities, but Mercouris chronicled how Britain had declined to cooperate with Russia. The British had rejected the Russian suggestion that the inquiry take place in Russia. They did not allow Russian investigators access to Boris Berezovsky, and refused to share their evidence of polonium poisoning. The case for Russian involvement in many ways was undermined by the inquiry: expert testimony demonstrated that polonium was widely available from many sources, not just from Russia, and was not nearly as expensive as initially thought. Evidence that either Lugovoi (who had been a KGB agent) or Kovtun were FSB agents was circumstantial, as was evidence of the FSB motive. The judge appeared unreasonably to credit Litvinenko's allegations against Putin, seeming to side with Berezovsky who, with the assistance of public relations firm Bell Pottinger, had promoted the narrative of Putin's responsibility. Owen attached weight to Litvinenko's "deathbed statement," even though this may have been authored by Litvinenko's colleague and ally of Berezovsky, Alexandar Goldbarb (who had also promoted the anti-Russian protests of punk rock group Pussy Riot). Litvinenko himself had previously accused an entirely different party, Mario Scaramella, of poisoning him. Finally, there were other suspects: people who might have wanted Litvinenko dead – one witness who interviewed Litvinenko on several occasions testified that he claimed to be blackmailing certain parties – and their motives were more compelling than any attributable to the FSB (Mercouris 2016; Dunkerley 2016).

## Movement towards lethal aid

The US House of Representatives passed a resolution (758) on December 4, 2014 titled "Strongly condemning the actions of the Russian Federation, under President Vladimir Putin, which has carried out a policy of aggression against neighboring countries aimed at political and economic domination." This was the kind of resolution, former US Congressman Ron Paul noted, that leads to sanctions and war (Paul 2014b). Without evidence, the resolution (paragraph 3) accused Russia of invading Ukraine, condemned its violation of Ukrainian sovereignty and (paragraph 11) demanded that Russia withdraw its forces. It urged Kiev to resume military operations against separatists in the eastern regions. In reality there was no unambiguous evidence of Russian troops in Ukraine (a necessary condition before any such troops be withdrawn). The call to Kiev for more aggression contravened the Minsk I ceasefire agreements and flew in the face of ongoing negotiations that would lead to Minsk II (Paul 2014b).

One week later, the US Senate passed Resolution S.2828, The Ukraine Freedom Support Act of 2014 that authorized lethal and non-lethal aid, building on previous Senate complaints, for which it had no unambiguous proof, of similar aid from Russia to the eastern separatists. These congressional developments marked a step beyond economic warfare – the Foreign Affairs Committee of the US House of Representatives was told in May that US sanctions were intended to damage Russian–EU business and destabilize Russia through currency instability and inflation – to military warfare (Martenson 2014).

In May 2015, the House Armed Services Committee passed the 2016 National Defense Authorization Act (NDAA) by 60 to 2 votes. Section 1532 of the 498-page document called for the USA to provide assistance, "including training, equipment, lethal weapons of a defensive nature, logistics support, supplies and services, and sustainment to the military and national security forces of Ukraine" through the end of September 2016. Already, the USA was training the right-wing Azov forces, alongside units of the Ukrainian National Guard near Lvivi (Cathcart and Epstein 2015). Canada, a close supporter of US Ukraine policy, furnished 200 Canadian Armed Forces (CAF) personnel to participate in the US-led mission to train the Ukrainian Army and National Guard in the Donbass region from September 2015. In November, US instructors initiated a six-month course for Ukrainian special operations forces in Khmelnytsky region. Simultaneously, the Pentagon announced it would upgrade its training commitment from a small mission for Ukraine's National Guard to a direct program for five battalions of Ukraine's regular military forces and special operations troops.

Ukraine's US Ambassador, Valeriy Chaly, stated in mid-2015 that Ukraine was actively armed by Western partners, including with lethal weapons. He alleged broad US support for providing more weapons. The USA, he said, was providing communications and night-vision systems, medical equipment, counter-battery radars, armored combat vehicles, robotic systems, including for mine clearance and imagery intelligence support. The Ukrainian President Poroshenko stated on Ukrainian state television in October that US President Obama had promised new models of defensive weapons, including counter-battery Q36 radar systems. The White House considered these a continuation of US provision of non-lethal support. The Ukrainian Interior Minister Arsen Avakov claimed on September 29 that Ukraine had already received US caliber sniper rifles. Aid extended to radar, medical supplies and communications gear, as well as economic assistance and military training totaling $265.5 million (Mazneva 2015).

The International Peacekeeping and Security Centre in Western Ukraine had developed into an important arena for joint maneuvers with NATO soldiers from 2013. From April to November 2015, 750 Ukrainian National Guard members were trained by 300 soldiers of the 173rd US Parachute Brigade. From November 2015 into 2016, 2,800 Ukrainian soldiers, 1,500 US soldiers and 1,000 soldiers from other countries participated in NATO's "Partnership for Peace" training program. In preparation, the Verkhovna Rada adopted a law allowing the presence of foreign troops in Ukraine for up to sixty-one days. The exercises made use of two mobile radars, type AN/TPQ-36 – each worth $15 million and gifted to President Poroshenko by the USA. Special equipment was supplied from US manufacturers working through Bulgarian intermediaries (Heyden 2015a).

US and NATO forces continued to amass military presence along borders shared between US-allied countries and Russia. In June 2015, the US Secretary of Defense, Ashton Carter, announced that the USA would begin to "preposition" large numbers of tanks and artillery along the Russian border to "reassure" its allies (Ditz 2015b). The USA engaged in military drills with Ukraine near its Polish border for the annual Rapid Trident exercise, amidst reports that Russia had flown jets over Baltic and Nordic countries and planned live rocket fire drills near Sebastopol (Gorshkov 2015). In December 2015, Patrick Auroy, NATO's Assistant Secretary General for Defense Investment, and Oleg Gladkovskyi, First Deputy Secretary of the National Security and Defense Council of Ukraine, signed off on the NATO-Ukraine Defence-Technical Co-operation to improve Ukraine's defense industry and armed forces, deepening cooperation between the alliance and Kiev in defense science and technology.

Early in July 2015, the Kiev regime agreed to purchase military weapons from the UAE, raising the suspicion that the US-led NATO alliance was establishing another backdoor into Ukraine to circumvent EU nervousness about lethal aid. Cunningham described the UAE as "largely a value-adding or marketing platform for established Western manufacturers," noting that the "UAE defense sector is dominated by US military imports and American weapons giants, Boeing, Lockheed Martin and Raytheon" (Cunningham 2015a).

## Putin as solely responsible for upholding Minsk II

US and Western powers, as at the conclusion of the G7 summit in Germany early in June 2015 pressured Putin and Russia, under threat of continued sanctions, to resolve the Ukraine crisis. Cunningham described this as "prove a negative" propaganda whereby the victim is pressured to show that he is NOT guilty of a false allegation, in this case that Russia was directly involved in and controlled Eastern Ukraine, in face of actual aggression by the NATO-supported Kiev government against the separatist provinces of Donetsk and Luhansk, and Kiev's refusal to negotiate with them (Cunningham 2015d). Parry chronicled WMM's portrayal of Putin as being single-handedly responsible for the situation in the Donbass – the only player responsible for ensuring compliance with Minsk II. He compared *New York Times*' editorializing with the "Two Minutes Hate" strategy for mass mobilization employed by elites in George Orwell's fictional *Oceania.* Parry called attention, as the *New York Times* did not, to Putin's own assessment:

> Today we primarily need to comply with all the agreements reached in Minsk [...] I would like to draw your attention and the attention of all our partners to the fact that we cannot do it unilaterally. We keep hearing the

> same thing, repeated like a mantra – that Russia should influence the southeast of Ukraine. We are. However, it is impossible to resolve the problem through our influence on the southeast alone.
>
> There has to be influence on the current official authorities in Kiev, which is something we cannot do. This is a road our Western partners have to take – those in Europe and America. Let us work together. [...] We believe that to resolve the situation we need to implement the Minsk agreements, as I said. The elements of a political settlement are key here. There are several.
>
> (Putin, cited by Parry 2015g)

Putin went on to identify the three most important elements of a political settlement as: greater autonomy or decentralization; special status for Luhansk and Donetsk; and amnesty. In apparent compliance with Minsk II, Ukraine's President Poroshenko announced early in July 2015 plans to decentralize power and provide a measure of autonomy to the Donbass. He described this as "vaccination against federalization," to head off the possibility that a more autonomous Donbass might block Ukraine's integration with Europe. Russia did not consider the degree of autonomy on offer was sufficient, nor did the leaders of ethnic Russians in Donetsk and Luhansk (Alexander Zakharchenko and Igor Plotnitsky, respectively) who claimed that neither had been consulted (Ogirenko 2015).

## Shooting down of the Russian civilian airliner over Egypt and fighter plane over Syria

Despite WMM outrage over MH17 in July 2014 – which, even in the absence of anything resembling adequate evidence, they quickly attributed to Russia or Russian-supported separatists, little sympathy was extended to the 224 victims of a Russian airliner downed over Egypt's Sinai Peninsula at the end of October 2015, nor for victims of Turkey's controversial shooting down of a Russian military jet on November 23, 2015 – on the likely false pretext that it had violated Turkish air space. On the first, Parry castigated the *Washington Post*, in particular, for: (1) "a stunning lack of sympathy"; (2) ridiculing Putin and the Egyptian President Abdel Fattah el-Sisi for their quite proper care not to rush to judgment that it was an act of terrorism; while (3) mocking the two leaders, unreasonably, for failing to vanquish the terrorists. Without evidence, the *Washington Post* linked the terrorist attack to "Mr. Putin's Syrian adventure" and hoped that it would inflict on Putin "a potentially grievous political wound" (Parry 2015f).

I have chronicled the intensification of WMM's demonization of Russian President Vladimir Putin and Russia generally. I have noted that demonization was: (1) frequently based on inaccurate presumptions (e.g., that Putin was a "dictator"); (2) projected on to Putin and his administration the aggressive words and acts of Western leaders themselves; (3) jumped to the most negative possible conclusions about the words and behaviors of Putin and his administration (e.g., allegations of Putin's "invasion" of Ukraine, Putin's "responsibility" for the murder of Nemtsov or the attack on MH17); (4) attributed to Putin a degree of control that he could not possibly have possessed (e.g., suggesting that he alone had responsibility for upholding Minsk II); and (5) frequently appeared motived by spite (quite out of kilter with the domestic popularity enjoyed by Putin), perhaps betraying fear of how Western publics felt about their own political classes.

# 8 The downing of MH17

A Malaysian Airlines Boeing 777–200ER, en route from Amsterdam to Kuala Lumpur, the Malaysian capital, on July 17, 2014, was flying at about 33,000 feet over separatist-held territory in southeastern Ukraine when it broke apart in mid-air and crashed outside of Hrabove, near Torez in Eastern Ukraine's Donetsk Oblast, killing all 298 passengers and crew members. Victims came from ten nations. Almost two-thirds, 193, were Dutch; forty-three were Malaysian; and twenty-eight were Australian.

The issues of who was responsible and, if a culpable party could be proven, the appropriate punishment, commanded much political and press attention. While full consensus as to the facts may never be achieved and controversy has persisted, the preponderance of evidence from Western sources appeared at the time of writing to support the contention that MH17 was bought down by a missile fired from a Russian-made BUK under the control of separatist forces and/or Russians who believed they were firing at a Ukrainian military jet. Because Russians and separatists were backed into a defensive position, notwithstanding disturbing questions as to why civilian aircraft were still flying over a war zone in which several planes had been attacked, Western leaders and WMM pressed home their propaganda advantage. The MH17 tragedy contributed towards the demonization of Putin and served to distract public attention from issues of the legitimacy of the February 2014 coup, the West's intervention, and the intensification of hostilities between nuclear powers.

## Initial BUK claims

Taking their cue from claims of the coup regime in Kiev – hardly an independent source – US politicians and WMM rushed to attribute the downing of MH17 to "pro-Russian" militants, possibly aided and abetted by Russians, who allegedly had used Russian weapons. The US Secretary of State, John Kerry, made the claim on August 12, 2014 that "We saw the take-off. We saw the trajectory. We saw the hit. We saw this airplane disappear from the radar screens. So there is really no mystery about where it came from and where these weapons have come from" (cited by Parry 2016b). This charge, long before the investigation, let alone proof (coming one year after comparable false confidence alleging Assad's use of chemical weapons in Syria) was used by the USA and EU to justify further sanctions against Russia, escalating a local to an international conflict, with nuclear potential.

The quality of emerging evidence 2014–2016 was mixed. WMM could have learned from a 2001 episode cautioning against their rush to judgment. Ukraine had shot down a Russian civilian airliner flying from Russia to Israel on October 4, killing seventy-eight people. Kiev denied the shoot-down for nine days, then acknowledged that it probably was responsible for "an accidental hit from an S-200 rocket fired during exercises" in Crimea. Kiev recanted this explanation in 2011, but offered no substitute (Boardman 2014).

Washington supported Kiev's claims that MH17 was shot down by one of possibly three Buk self-propelled fire installations (BUK 312), of Russian origin allegedly manned by the pro-Russian Eastern Ukraine resistance. One source attributed responsibility to a BUK 312, supposedly located in rebel-held Snizhne. The "evidence" relied on a Ukraine government audiotape in which resistance commanders acknowledged responsibility for the shoot down. A key part of the recording appeared to have been made hours before the crash. Additionally, the tape was patched together, indicating that the conversation in question was not continuous. The White House position, announced only two days after the incident, was contradicted days later by a "government assessment" from the Director of National Intelligence released on July 22, 2014. This listed other weapons allegedly provided by Russia to ethnic Russian rebels yet made no mention of even a single Buk anti-aircraft missile system which, had it been there, would hardly have been overlooked by US satellite surveillance (Parry 2015j). Not until February 2016 did Dutch prosecutors confess they had no satellite images allegedly because of cloud cover on the day in question (although radar satellites working at microwave frequencies can see through clouds). They conceded there was radar evidence which, mysteriously, they were still in the process of acquiring although for "unknown reasons" Ukraine had not been running primary radar surveillance that day, even though there were no other means for ensuring air safety over the war zone in Eastern Ukraine (RT 2016b). No other means, that is, besides shutting down the air space.

Evidence emerged of other separatist and Russian commentary in which speakers appeared to believe either that a Ukrainian military jet had been attacked, or that a civilian airliner had been downed in error. The Donetsk separatist leader, Igor Strelkov, reportedly uploaded a report on the Russian social media site, VKontakte, to celebrate the downing of what he thought was a Ukrainian AN-26 jet. This posting was said to have been removed subsequently. A pro-American source (Fitzpatrick 2014b) cited similar "confessions" in Russian media, notably in the pro-Kremlin newspaper *Vzglyad*. These indicated that separatists had the means to bring down Ukrainian jets by July 14, means that likely included one or more BUKs either provided by Russia or stolen from Ukrainian forces. Another possibility was that missiles were fired from Russia. The independence of such evidence was uncertain. Alexander Borodai, Prime Minister of the Donetsk People's Republic, was reported to have called a

Russian media contact forty-nine minutes after the crash to say that separatists had likely shot down a civilian airliner. As with the articles from *Vzglyad*, above, this evidence came from *The Interpreter* by Catherine Fitzpatrick (2014a) which in turn was backed by the US propagandist organization Radio Free Europe/Radio Liberty. The article acknowledged that the interviewee Dmitry Muratov, editor of Moscow opposition newspaper *Novaya Gazeta*, considered it likely that Borodai had indeed made the statement attributed to him, but that Borodai himself denied he had made such a claim. Within days of the attack his spokesman denied that rebels had the equipment to shoot down a plane flying as high as MH17. It was scarcely surprising that noted anti-Putin intellectuals such as Muratov, or politicians like Duma member Ilya Ponomarev, should side with Ukrainian and Western versions of events.

Perhaps greater weight could be given to an Associated Press (AP) story on July 25 claiming that a BUK missile launcher actually stopped in front of AP journalists in Snizhne on its way East, where "a man wearing unfamiliar fatigues, speaking with a distinctive Russian accent, checked to make sure they weren't filming" (Karmanau and Leonard 2014). The story cited an unnamed "rebel" as saying later that the separatists shot down a civilian airliner in error. It was a remarkable coincidence that AP had journalists in Snizhne, and that the BUK launcher pulled up directly in front of them. These journalists do not appear to be authors of the news report, and the report was written over a week after the shooting. "Unfamiliar fatigues" do not have to be Russian, and a "distinctive" Russian accent would not be strange among ethnic Russian speakers of the Donbass close to the Russian border. There was an earlier AP story, on July 17, which asserted that an AP reporter (unidentified, presumably one of the AP reporters cited in the July 25 story) had seen a BUK launcher in Snizhne (Leonard 2014). If true, the reporter or reporters were possibly in Snizhne investigating shootings on Ukrainian Sukhoi-25 planes either by rebels or from Russian territory, and exploring Ukrainian allegations of sophisticated equipment being supplied to the rebels from Russia.

A disturbing implication emerges. On the one hand, there were apparently multiple reports from Ukrainian sources before, on and after July 17, claiming knowledge of sophisticated Russian weaponry in the hands of either or both Russians and ethnic Russian separatists, possibly including one or more Buk missile launchers, as well as reports that Russia was itself deploying Russian equipment from within Russian territory in support of ethnic Russian separatists. There was even a report from one of the world's major news agencies, the American agency AP, specifically claiming a direct AP sighting of a Buk missile launcher in Snizhne, possibly with Russians amongst its crew, heading east on the morning of July 17, 2014. This AP report was datelined 10.31 a.m., July 17. The local time of the MH17 crash in Eastern Ukraine was 13.20 p.m. that day. Since Ukraine is thirteen hours ahead of US east coast time, the AP report as it appeared on ABC's site was filed no later than thirteen hours ahead of 10.31 a.m. of July 17 – namely, 9.31 p.m. on July 16.

Why is this significant? Four months earlier, some international airlines decided not to fly over Eastern Ukrainian airspace. These included Korean Air,

Asiana Airlines and British Airways. The International Civil Aviation Organization warned governments in April of the risk to commercial airliners flying over Ukraine, and the American Federal Aviation Administration advised "extreme caution." If there was any reason for Ukraine to keep open its airspace in Eastern Ukraine, or for airlines to continue to fly over it, it would be the tenuous proposition that civilian airlines are generally required to maintain altitudes of above 30,000 feet and that ethnic Russian separatists did not have the firepower to reach such altitudes. Both Ukrainian and, across the border, Russian military had access to numerous BUK missile systems, and Ukrainian military had many such systems located in contested areas of Eastern Ukraine. In other words, there was never at any point during the fighting a time that civilian airlines were not at grave risk of being fired upon, whether by accident or for any other reason, from BUK missile systems with a reach of up to 70,000 feet. There had been a growing rumble of concern well before July 17, from both Ukrainian sources and WMM, of successful attacks on Ukrainian military jets by ethnic Russian separatists, amidst allegations of the deployment of sophisticated Russian weaponry in Eastern Ukraine or the support of such weaponry from the Russian side of the border and indications – even visual sightings – of Buk missile launchers. In these circumstances, it is difficult to understand why Ukraine had not shut down its airspace over Eastern Ukraine before July 17 or, given that Ukraine had not shut down its airspace, why international airlines did not simply divert their flight paths to avoid the obvious and imminent dangers that were so clearly present. These considerations suggest possible criminal negligence, at best, and even something worse, on the part of Ukrainian authorities and international aviators regardless of who was actually responsible for the shooting down of MH17.

The same considerations invite other questions. Could it be that Ukrainian authorities simply did not believe their own propaganda about equipment allegedly deployed by the Russians in support of or handed over to ethnic Russian separatists? If they did believe it, there was no conceivable excuse for not shutting down Eastern Ukrainian airspace. If they did not believe their own propaganda, why did the Ukrainians have their own BUKs in eastern Donetsk – who were they going to fire at?

Perhaps Kiev authorities did not believe their own propaganda because they reasoned there might be some plausible scenario in which they would need to fire their BUKs. Ukrainians and WMM had made much of the possibility of a Russian "invasion." They could not predict when exactly such an invasion might take place, or if the feared invasion might simply take the form of penetration of Ukrainian airspace by Russian fighter planes. So why did not the Ukrainians, as a responsible and necessary precaution, shut down their airspace to civilian flights anyway?

Alternatively, did Kiev believe its own propaganda and calculate that eventually an international civilian airliner would indeed be shot down by separatists or Russians (especially if Ukrainian jets routinely used civilian airliners as camouflage, as one source has alleged – see below) and that this would provide Kiev a major propaganda coup? Or did Kiev plan on shooting down such an airliner and then to blame the separatists or Russians or both – a high-risk strategy given US satellite surveillance, unless the USA was party to any such conspiracy? Or did civilian airliners seem so valuable as camouflage for Ukrainian military planes that the Ukrainian authorities decided to keep the skies open just for that purpose?

## Russian-sourced alternatives

Significant alternative accounts later materialized. If the BUK 312 had been located in rebel-held Snizhne, why had there been no BUK defense of Snizhne just two days before the MH17 shoot down when Ukraine's air force attacked the city? A source interviewed by the Russian investigative journalist Anatoly Sharij (Collins 2014c) was a 23-year-old, former BUK 312 crewmember who claimed that the missile unit had been manned by the Ukraine military, it was 100 percent Ukrainian made, and there was only a slim chance that the crew had shot down MH17. Air offensive crews would have expected to aim at targets flying far lower than the regulated minimum 30,000–35,000 feet for civilian airliners. Whichever unit shot it down (other narratives would discount the role of a BUK missile altogether) would likely have been Ukrainian, and would have known that the aircraft could not have belonged to the resistance for the obvious reason that the resistance had no air force. They might conceivably have thought it was a Russian fighter jet – even the airliner carrying the Russian President Vladimir Putin over a different part of Ukraine at this time.

Following the Sharij interview in December 2014, Russia's MH17 Investigative Committee confirmed the claims of a Ukrainian airbase worker that on the day of the attack he witnessed the departure of a Ukrainian air force Su-25 combat jet from an airbase in eastern Dnipropetrovsk carrying air-to-air missiles – and returned without them. This supported the single most important alternative to the BUK narrative, namely, that MH17 was hit by machine-gun fire from a Ukrainian fighter jet acting either deliberately or in error. According to London's *Daily Mail* on December 5, a Russian journalist's video documentary "suggested" that pieces of 30 mm rounds, the size of bullets from types of fighter-jet planes of the Ukrainian Air Force, were found in the pilots' bodies (Zuesse 2014e). Nonetheless, the final report of the Dutch Safety Board investigation, along with other sources, firmly rejected the air-to-air missile theory.

## Washington's account

Washington's political establishment seized on the narrative of pro-Russian culpability; the US intelligence establishment was less outspoken. Senator Saxby Chambliss (Republican, Georgia), member of the Senate Intelligence Committee and privy to all the relevant evidence seen by the State Department, acknowledged that while there was some "circumstantial" evidence pointing to Russia, he saw no smoking gun (Ditz 2014a). WMM circumscribed their comments within a corral of militant neo-conservative sources, provoking one experienced Dutch journalist, Karel Van Wolferen, to lament:

> The Dutch media also appeared unaware that the catastrophe was immediately turned into a political football for White House and State Department purposes. The likelihood that Putin was right when he said that the catastrophe would not have happened if his insistence on a cease-fire had been accepted, was not entertained.
>
> (Van Wolferen 2014)

Evidence normally available for public consumption in these circumstances would include transcripts of conversations between pilots of the ill-fated aircraft and the nearest Air Traffic Control tower. These still had not been disclosed many months later (Maavak 2014), although they are referenced in the final report of the Dutch Safety Board in late 2015. Ukraine's SBU intelligence agency issued a video that purported, as stated above, to feature an intercepted call of resistance fighters discussing the shoot-down. There is reference to this in Koshiw's book (2015, see below). The video carried a creation time-stamp suggesting it was created a day before MH17 was shot down. The conversation may have been a "doctored" version of an intercept related to the recent downing by resistance fighters of Ukrainian military transport planes (Ditz 2014a).

The US government nonetheless asserted that an SA-11 surface-to-air missile had been fired by separatists, arguing that the BUK launcher apparently used to shoot down the plane was "Russian made," and imported into Ukraine from Russia. That it might have been "Russian made" could hardly have been a surprise. Nearly all Ukrainian military weaponry at the time was "Russian made." Ukraine was part of the Soviet Union until 1991 and continued to use mostly Russian military equipment (Parry 2014s). Yet much of this "Russian" equipment used by both the Ukrainian army and separatists was actually manufactured in Ukraine under contract to Russian armed forces. For separatists to have had access to a BUK missile system did not require any "importation" from Russia, merely access to equipment already stationed in Ukraine.

Not all US military personnel endorsed Washington's talking points. A *Washington Post* report cited Air Force General Philip M. Breedlove, US commander of NATO forces in Europe, declaring that "We have not seen any of the [Russian] air-defense vehicles across the border yet," while Rear Admiral John Kirby, the Pentagon press secretary, offered that "defense officials could not point to specific evidence that an SA-11 surface-to-air missile system had been transported from Russia into eastern Ukraine" (cited by Parry 2014s). Pro-Russian separatists denied they had missiles with the reach of the 33,000-foot altitude of the Malaysian airliner. Russian sources indicated that Ukrainian military had positioned BUK anti-aircraft systems in Eastern Ukraine and that the radar of one battery was active on the day of the crash. The Russian Defense Ministry stated that Russia had detected the activity of a Kupol radar on the day of the attack, deployed as part of a BUK-M1 battery near Styla village,

30 kilometers south of Donetsk. Buk missile systems are large and must be transported on trucks, and it would be difficult to conceal their presence from US aerial surveillance which had been concentrating intensely on the Ukraine–Russia border in recent months.

How did the USA ascertain that the missile was an SA-11 as opposed, say, to other versions of the Buk missile system? Because that was what the Kiev government had claimed? Former congressman Ron Paul unraveled a logic that mainstream media could not: (1) The Kiev government, he noted, also used Russian weapons and was at least equally as likely to be the culprit; (2) the coup in Kiev had been instigated by the USA/EU in the first place; (3) the coup government had already killed 250 people in the breakaway Luhansk region, June to September, in attacks that mainly killed citizens; (4) neither the separatists nor Russia had anything to gain and everything to lose from the bombing of an international passenger jet, whereas the Ukraine had much to gain from dissemination of evidence purporting to show that Russians or ethnic Russian Eastern Ukrainians were responsible; and (5) the missile in question was part of a sophisticated system that required training which the separatists had not received (Paul 2014a).

Meanwhile, unnamed CIA analysts cited US satellite reconnaissance photos as evidence that the anti-aircraft missile was actually fired by Ukrainian troops from a government battery, not by ethnic Russian rebels (Parry 2014p, 2014n). Soldiers manning the battery wore Ukrainian uniforms and may have been drinking since beer bottles were scattered around the site. Official US spokespersons around this time promoted the idea that a Ukrainian "defector" may have fired the missile. This notion was interpreted as an attempt to get around the embarrassment (to the Kiev government and the USA) occasioned by the presence of Ukrainian uniforms around the battery. Suspiciously convenient opposition evidence was presented by a Reuters report citing "a powerful Ukrainian rebel leader" who "confirmed that pro-Russian separatists had an anti-aircraft missile of the type Washington says was used to shoot down Malaysia Airlines flight MH17 and it could have originated in Russia" (Zverev 2014).

The official narrative begged many questions. Why was a flight from Amsterdam to Malaysia directed over a war zone? Why would Ukrainian authorities have approved it? The plane's normal route had been diverted to avoid bad weather (Parry 2014s) by about 200 miles to the north, over the Donetsk war zone (Boardman 2014). When the pilot asked to fly at 35,000 feet, air controllers reportedly ordered him to fly at 33,000 feet. If Russia had provided BUK missiles with a range of up to 70,000 feet, as known to both Kiev and Washington beforehand, why was MH17 ordered to fly within their range? There were questions about procedures at Amsterdam's Schiphol airport, where security was operated by ICTS, an Israeli company founded by former officers of Israeli's Shin Bet intelligence agency, and there was an apparently unexplained presence of "foreign" advisors in Kiev's control tower.

## Fired at from the air

The BUK narrative was challenged by opposing claims from the Russian Ministry of Defense that a Ukrainian SU-25 (some sources said two) had trailed MH17 (Escobar 2014b). The damaged MH17 starboard jet engine indicated a hit from an air-to-air missile, not a BUK. Not a single eyewitness had seen the graphic, thick, missile trace that would have been visible had a BUK been responsible (Escobar 2014b). This account was corroborated by a BBC report citing eyewitness accounts of villagers who had seen a fighter close to the airliner around the time of its crash and heard explosions coming from the sky. The BBC later removed the report from its archive (Whitney 2014g) but resuscitated its account for a 2016 documentary (Rudin 2016). Michael Bociurkiw, among the first OSCE inspectors to reach the crash site spent over a week examining the wreckage and described to CBC World News two or three "really pock-marked" pieces of fuselage. "It almost looks like machine gun fire; very, very strong machine gun fire that has left these unique marks that we haven't seen anywhere else" (cited by Collins 2014a).

Despite Western sanctions and denunciations, a columnist for the *World Socialist Web Site* noted:

> Washington has presented not one shred of evidence that Flight MH17 was brought down by a missile either fired by the anti-Kiev forces or supplied by Moscow. Instead, it has pointed to videos and audio recordings posted on YouTube by the Ukrainian intelligence agency, which is notorious for fabricating false charges against Russia. The vast US intelligence apparatus has produced no satellite photos, communications intercepts or anything else to back the allegations against the Putin government.
>
> (Van Auken 2014b)

By contrast, the Russian military, had:

> released radar and other intelligence data raising significant questions about the Western narrative. They show that the pro-Kiev forces had at least three anti-aircraft batteries in the area on the day of the crash, and that a Ukrainian warplane armed with air-to-air missiles had flown to within striking distance of the Malaysian airliner shortly before the crash.
>
> (Van Auken 2014b)

This view was endorsed by the veteran investigative journalist, Robert Parry, who complained that the US Administration had not supplied any direct evidence, only Ukrainian intelligence claims (Parry 2014n). The Security Service of Ukraine (SBU) was headed at the time by Valentyn Nalyvaichenko, described as a known CIA agent by "The Saker" [Whitney 2014h]). There were social media snippets allegedly of a BUK system being transferred by Russians to rebel control, shipped into Ukraine or deployed into firing position and then being taken back into Russia. But the continuing absence of evidence from the USA, whose spy satellites were positioned over Eastern Ukraine and who had released previous images of Ukraine taken by such satellites, indicated bad faith. Not enough to prevent the *New York Times* from assuming Russia's culpability as an incontrovertible fact nor Washington from leveraging what it alleged to be Russian responsibility as the pretext for intensifying sanctions (Parry 2014c).

## Ukraine's role

If Russia or ethnic Russian rebels were not responsible, then the chief remaining suspect would be the Ukrainian government which was known without doubt to be in possession of BUK anti-aircraft missiles and had reportedly (see above) sent two fighter jets into the vicinity of MH17 at the time of the shoot-down. Citing independent analyses of the initial evidence from the crash site, Parry considered that the jetliner may have been destroyed by an air-to-air attack, not by an anti-aircraft missile fired from the ground, or that: "a Ukrainian military BUK battery and the jetfighters may have been operating in collusion as they hunted what they thought was a Russian airliner, possibly even the plane carrying President Vladimir Putin on a return trip from South America" (Parry 2014n). Anonymous sources cited by the Interfax news agency and in turn cited by Roberts (2014c) claimed that the MH17 and Putin's airliner were traveling on an almost identical route within a few minutes of one another (this appears to be discounted by the final Dutch Safety Board report, which says Putin's airliner had been diverted well north). A fire started by the Ukrainian forces at the site of the crash threatened to contaminate surviving evidence of shrapnel holes that were thought by independent analysts to be possible evidence of an attack by Ukrainian jetfighters.

Western intelligence releases appeared to stall after August 2014. The Dutch coordinator for the struggle against terrorism claimed that Dutch authorities already had extensive data from the black boxes and other sources in their possession. Yet Russian authorities complained that the plane's black box recordings had not been made available, that Kiev persistently refused to publish the recordings of the conversations between the MH17 pilot and Ukrainian air traffic controllers, and that no information was forthcoming from the International Civil Aviation Organization, the United Nations aviation agency. The Russian foreign minister, Sergey Lavrov, warned:

> We must not allow the investigation of the MH17 crash to be manipulated into oblivion as already happened to investigations of many Ukrainian tragedies, including the sniper assault against civilians in Kiev in February, massacres in Odessa and Mariupol in May, and others.
>
> (Lebedev 2014)

## Preliminary report

Western reports of the preliminary results of the Dutch government Air Safety Board investigation appeared on September 8, 2014. According to the Convention on International Civil Aviation, the country in which an aviation incident occurs is responsible for investigating the incident but may delegate that responsibility to another state. The main team was led by the Dutch. Members included twenty-four investigators from Ukraine, Malaysia, Australia, Germany, the United States, the United Kingdom and Russia. The investigating team issued a call for witnesses, particularly with respect to the movement of a BUK missile launcher from Sjevernyi near the Russian border through or near Luhansk to Donetsk on a Volvo truck. Also investigating were a Malaysian team of 133 officials and experts, an Australian forty-five-member panel, six investigators from the United Kingdom and extra consular staff from the UK Foreign Office.

Early WMM articles on the preliminary report cited evidence that "appeared to confirm initial assertions that the passenger plane was hit by a surface-to-air-missile in mid-flight July 17 before crashing in Ukraine." Accounts of the report referenced "impacts from a large number of high-energy objects from outside the aircraft." Aircraft fragments revealed numerous puncture holes and indentations on the plane's skin that the investigators said would be consistent with damage from missile shrapnel and that ruled out pilot error or mechanical fault as the cause of the disaster. The report referenced evidence of cockpit voice and flight data recorders as giving no indication of any trouble with the flight (Werth and Williams 2014).

Yet the report answered few critical questions, evading even some that would seem relatively easy to address – such as the veracity of Russian military radar evidence showing a Ukrainian SU-25 jetfighter in the area – a claim that Kiev denied (Parry 2015h). The report did not discuss whether the US government had satellite surveillance evidence that would reveal exactly where the supposed ground-to-air missile was launched and who may have fired it. Nowhere in the report could be found "evidence of satellite images of missile launchers, intelligence from the USA regarding missile launches, or any information or evidence at all in any regard suggesting a missile had destroyed MH17" (Cartalucci 2014a), even though comparable evidence was available in abundance a year later when Russia sent air support to attack ISIS units in Syria. The absence of input from the USA, Cartalucci observed, suggested that the USA had nothing to offer in its support of its anti-separatist narrative, hinting instead that NATO might itself have been the culpable agent.

Curiously the preliminary report made reference only to post-crash satellite photos, even though Russian military released satellite images that purportedly revealed Ukrainian government BUK missile systems positioned north of the eastern Ukrainian city of Donetsk prior to the attack. These included two batteries that allegedly were shifted 50 kilometers south of Donetsk on July 17, the day of the crash, and then removed by July 18. Analysis by Bellingcat.com (with which there are problems) compared the Russian satellite results with images of the same locations using Google Earth satellite photos, concluding that the Russian photos were taken a month prior to the attack, and that the Russians had used Adobe Photoshop to edit and compress the images (Bodner 2015b). Some US intelligence analysts shifted away from the original BUK narrative that blamed the rebels and Russia, to consider whether extremist, pro-Kiev elements were responsible. The preliminary report did not address evidence regarding who actually fired the missile that was supposedly responsible, or from precisely what location (Parry 2014s).

## German intelligence

In contrast to Dutch Safety Board (DSB) uncertainty, the German intelligence agency (BND) determined in October that MH17 had been shot by surface-to-air missiles fired by pro-Russian separatists using a BUK that separatists had acquired from Ukrainian government stockpiles. In an interview with *Der Spiegel*, DSB head Fred Westerbeke (also chief investigator with the Dutch National Prosecutors' Office) seemed less than impressed, saying that his team was still (!) awaiting the arrival of satellite images from the USA (Leurs 2014). An anonymous bounty of $30 million for information as to who shot down MH17, who gave the order, and who was covering up, appeared unfruitful. Westerbeke allowed for the possibility that the aircraft might have been attacked from the air, *Der Spiegel* reported. The German expert Peter Haisenko estimated that MH17 was hit by the type of double-barreled 30-mm guns used on Ukrainian SU-25 fighter planes (Whitney 2014g) – a proposition for which, as we have seen, a BBC report had cited witnesses. The report had been captured and uploaded to YouTube before being removed from the BBC archive. Global Research posted a transcript on September 10, headlined, "Deleted BBC Report. 'Ukrainian Fighter Jet Shot Down MHI7,' Donetsk Eyewitnesses" (Zuesse 2014c).

Russia claimed it had tracked the Ukrainian fighters on radar. Tweets from a worker at Kiev's Air Traffic Control at Borispol confirmed this. Shortly after the tweets, the Ukrainian military reportedly took possession of the control tower, seized air traffic's recordings, and the worker disappeared. Continuing Western secrecy regarding the black boxes, US satellite images and Ukrainian air traffic control radar recordings fed suspicion of cover-up (Zuesse 2014c), extending to the coroner's report on the pilot's corpse. Bodies of both main and support crews were in the Netherlands long after bodies of passengers were released to relatives. Zuesse speculated that the pilot's body was riddled with bullets, as would be the case had the plane been attacked by machine gun fire from another aircraft (Zuesse 2014e). Supporting evidence for this theory was provided by the *Daily Mail* on December 5, cited by Zuesse (2014e), which referred to a video documentary from the Russian journalist, Andrei Karaulov, claiming there was evidence that "pieces of 30mm rounds were found in the bodies of the pilots." Such bullets are the same size as those from types of fighter-jet planes used in the Ukrainian Air Force, including Su-25, Su-27 and Mig-29. They are very different from the missile shrapnel which the USA and Ukraine alleged had brought down MH17. Since separatists had no aircraft, the presence of bullets would point towards a Ukrainian air attack.

The progress of the Dutch investigation was frustratingly slow for victims' families. Some German families planned to sue the Ukrainian authorities in the

European Court of Human Rights for Ukrainian incompetence for not shutting down the air space over which it had insufficient control. Another group of relatives requested that the UN take over the investigation and blamed the Dutch for what the group said was its mishandling of inquiries into the disaster.

## Criminal investigation

In addition to the enquiry of the Security Council of the Netherlands (Dutch Safety Board), a Dutch criminal investigation by the Joint International Criminal Investigation Team (JIT) under the direction of the Public Prosecution Service of the Dutch Ministry of Justice comprised prosecutors and police from Australia, Belgium, Malaysia and Ukraine. All told, it involved dozens of prosecutors and 200 investigators. Because the Kiev government could be considered a suspect, some critics opined that Ukraine should never have been handed the power conferred by its membership of the JIT to veto the publication of findings. Ukraine's inclusion, said New York-based geopolitical analyst Ulson Gunnar, writing in *New Eastern Outlook* magazine appeared a monumental conflict of interest, given that Ukraine was known to have had weapons capable of reaching the altitude that MH17 was flying when hit (Gunnar 2014).

The cooperation of implicated parties was less than 100 percent. MH17's black boxes were obtained only as a result of back-channel talks between the Malaysian Prime Minister Najib Razak and pro-Russian separatist forces in Eastern Ukraine where MH17 crashed. Razak also negotiated for the separatists to recover the bodies and allow access to the crash site. Kiev suppressed information regarding extensive contacts between the Kiev airport air traffic control and MH17. Months later, US and German intelligence had still not provided satellite images backing up their claims to have definitive proof of Russian involvement. The Dutch cabinet apparently ruled out the possibility that the investigation would also examine evidence that the plane might have been shot down in the air by a fighter jet, probably Ukrainian, known to have been in the area at the time.

Russia, an interested party like Ukraine, was excluded as a principal member of the criminal investigation. Even more surprisingly, Malaysia was (initially) excluded – on account of its "political neutrality" according to some media reports. This was astounding on several accounts: notably its implication that "neutrality" could be anything but an asset in such a matter. Belgium – not the EU – had barely anything to do with the flight or its fate, yet was included in the investigation. Malaysia owned the airplane in question. The Malaysian police chief and its transport minister expressly asked to be involved. Further, the International Civil Aviation Organization (ICAO) Chicago Convention, requires that the operator of a crashed flight must be included in an investigation. Malaysia was finally accepted as a full member in late November 2014.

The Director of Malaysia's Suaram human rights organization, Kua Kia Soong, criticized a non-disclosure agreement between the (original) four participating countries. According to Dutch state radio NOS, "Part of the agreement is that signatories retain control of the information they contribute. So they can veto disclosure of their own data" (Lantier 2014d). The Dutch Prime Minister Mark Rutte rejected Freedom of Information Act (WOB) requests by Elsevier into seventeen documents relating to the JIT pact, asserting that the benefits of disclosing information were outweighed by the risk of damage to Dutch relations to other states and world bodies. While each state was at liberty to select what information it would present to the joint investigation, members wanted to keep this information, even the terms under which it was being processed, hidden from the public (Lantier 2014d).

## German Correct!V investigation

The Russian foreign minister, Sergey Lavrov, was reported by RT in mid-January 2015 as complaining that the West appeared to have forgotten about MH17. He hoped that the Dutch would soon release their preliminary findings. Within days there appeared a report from an apparently independent, journalistic source, Germany's Correct!V. On its website, Correct!V described itself as the first non-profit investigative newsroom in the German-speaking world, established by the Brost Foundation, an entity associated with publishing interests and financed by charities and partner news media. The report confirmed some basic details of original US accounts – so conveniently that the symmetry in itself raised suspicions. It concluded that MH17 was downed by a ground-launched BUK missile, fired by the 53rd Russian Air Defense Brigade from Kursk, a unit that was protecting Russian tanks operating in the Ukraine that did not display national emblems. It also blamed Ukraine for exploiting civilian aircraft as a cover to protect Ukrainian fighter jets from BUK-missile tracking units. It blamed the EU for not issuing stronger warnings to Ukraine and to civilian airlines, and blamed the airlines for continuing to fly over a dangerous conflict zone. Oddly, it did not blame Ukraine for not shutting down its own airspace. Ukraine had a financial motive for not sealing its airspace: Ukraine received overflight fees for every international commercial aircraft that flew over its territory.

Correct!V noted that within thirty minutes after the plane crashed, the commander-in-chief of pro-Russian separatists in Donbass, Igor Girkin (previously referenced as Igor Strelkov) allegedly used social media to celebrate what he claimed was the downing of a Ukrainian military jet. The post was removed and Girkin later declared it false. Correct!V cited an anonymous expert source – who might well have worked for US/NATO agencies, and whose view was backed by a named Dutch fighter pilot trained by USAF – confirming that MH17 had been shot by surface-to-air missile. Three reasons were given: only a missile would have had the necessary force; missiles are programmed to explode just in front of or above the targeted plane, which forces the target to fly through thousands of pieces of deadly shrapnel; only a BUK missile would had the necessary reach. BUK missiles were Soviet developed and operated by radar. Radar cannot identify the type of airplane that is targeted. It can alert the targeted aircraft allowing it to pre-empt an attack: hence there is grave danger for both attacker and target. The missile launchers must strike while the target is still at high altitude.

The report cited the US State Department claims of June 2015 that Russian tanks had been spotted near the Ukrainian town of Snizhne, and referred to NATO-published photos of such tanks. It noted that Ukrainian forces had escalated their air attacks in the weeks prior to the MH17 incident, losing four aircraft in the process, of which some may have been attacked by BUK missiles. (This accentuates the mystery as to why Ukraine did not shut down Eastern Ukrainian airspace to civilian airliners). Hundreds of civilian planes were still flying over the zone at levels of over 10,000 feet. The presence on the ground of Buk missile launchers capable of hitting such targets and whose radar betrayed their presence gave targets just enough time, if fast, to take evasive action. Launch teams were highly motivated to shoot while the target was at high altitude so as to avoid being fired upon themselves. Because Ukrainian jet fighters used civilian airliners as "human shields," the EU, Ukraine and NATO should have issued strong warning to airlines, and airlines should in any case have stopped flying over such a dangerous zone.

The report disparaged the Russian claim that there was a BUK unit near the Ukrainian-controlled village of Zaroshchenske, finding no witnesses there to the presence of a BUK unit and that the territory in question had been under separatist control. Russian culpability was suggested by: (1) multiple authenticated photos (Bellingcat.com is cited as an important source) of a BUK launch pad near Snizhne that had been dispatched by Russia's 53rd Air Defense Brigade, based in Kursk; (2) local separatist confirmation of the presence of Russian tanks; (3) witness claims to having seen a plane shot down in that area; (4) one witness who asserted that the missile was fired by Russian soldiers; (5) video evidence of a BUK launcher returning to Russia with one of its missiles missing; and (6) an authoritative separatist source who denied that separatists had the expertise to successfully use a BUK system.

There were some troubling features of the report. Its narrative aligned perhaps too conveniently with US interests: not only did it ascribe primary responsibility to the Russians, but also lent encouragement to the interpretation that Russians deliberately shot down MH17, whereas much of the cited evidence underlined the scope for accidental judgment in the use of a BUK launcher. It is very unclear what Russia's motive for such an attack might be. The report discredited the idea that the plane was shot down by Ukrainian fighter jets, indicating that the impact of a BUK missile would leave traces not dissimilar to those that might be left by shrapnel from a fighter jet (although that does not disprove the possibility of shrapnel). It offered an explanation for the presence of Ukrainian jet fighters close by, namely, that Ukraine used civilian aircraft as "human shields." And it endorsed US, NATO and Ukrainian claims of direct Russian military involvement on Ukrainian territory which, as we have seen, was contested by the Russians.

The report relied principally on pro-US and pro-NATO sources. Many field witnesses cited were anonymous, perhaps to protect them against retaliation. The report's tone towards Russia and Putin was generally negative. Its dependence on Bellingcat.com (like later television investigations from Australia and Britain) is substantial. Bellingcat.com presented itself as principally a one-man operation (Eliot Higgins, operating from a Leicester, UK address).

Higgins appeared to work closely with Dan Kaszeta, a consultant formerly of the US Army Chemical Corp. These men were used as key sources for stories in the *New York Times*, the *New Yorker* and HRW that promoted arguments to establish Syrian regime culpability for the chemical weapons attacks on Ghouta in the summer of 2013. Their arguments were based on missile direction, missile range, the reported presence of hexamine in soil and other residues, and the view that only the Syrian government would have been capable of such a missile attack. Each claim was refuted by Lloyd and Postol (2014b; see also Schofield 2014) who concentrated on the area of the attacks in the eastern suburb of Zamalka, where the largest quantity of sarin was collected. They found that: the missile direction was from the north not the northwest; the improvised missile had a range far too short – about 1.2 miles, where a range of 3.5 to 6 miles would have been necessary to have hit East Ghouta from Syrian government positions; there was no scientific source for the idea that hexamine is minimally suitable as an acid neutralizer for the making of sarin bombs; and rebel forces would have been perfectly capable of constructing the missiles in question – all they needed was access to a machine shop with modest capabilities. The USA claims to possess satellite images showing the impact points of the chemical weapons were unlikely to have been reliable given that the charges that detonate chemical weapons are generally too small for the impacts to be visible by satellite.

Of the Syrian case, Lloyd and Postol noted that although widely quoted as an expert in WMM, Higgins "has changed his facts every time new technical information has challenged his conclusion that the Syrian government must have been responsible for the sarin attack" (Lloyd and Postol 2014a). Higgins claimed to monitor over 450 YouTube channels daily, looking for images of weapons and tracking when new types appeared in a war, where and with whom. He then matched such video evidence against Google Earth images to find exact locations (a process that may be open to interpretation on the grounds of the ambiguous visual evidence). Higgins was among the first to report the alleged use of improvised barrel bombs by the Syrian government, the alleged use of cluster bombs in 2012, the proliferation of shoulder-launched heat-seeking missiles known as MANPADS; and the proliferation of Croatian-made weapons. It is unclear what motives directed his investigations and to what extent his reports were themselves fact-checked.

Higgins' name was exploited by significant players including NATO's General Breedlove who tweeted a link to a May 27 story in the *New York Times*. Under the headline: "Armed With Google and YouTube, Analysts Gauge Russia's Presence in Ukraine" he referenced a report (*Hiding in Plain Sight: Putin's War in Ukraine*) from the Atlantic Council, a Cold War style think-tank. The report's leading analyst was Eliot Higgins (Boardman 2015a). The co-author of the report was John Herbst, a former US ambassador to Ukraine. The ex-bureau chief for the *International Herald Tribune*, Patrick Smith, commented wryly:

> Manipulating social media "evidence" has been a parlor game in Kiev; Washington; Langley, Virginia, and at NATO since the Ukraine crisis broke open. Look at the graphics included in the presentation. I do not think technical expertise is required to see that these images prove what all others offered as evidence since last year prove: nothing. It looks like the usual hocus-pocus. [...] Examine the Bellingcat web site and try to figure out who runs it. I tried the about page and it was blank. The site consists of badly supported anti-Russian "reports" – no "investigation" aimed in any other direction.
>
> (Smith 2015b)

The investigative journalist, Robert Parry, was bluntly dismissive of Higgins. He noted that Bellingcat collaborated with the USAID-funded Organized Crime and Corruption Reporting Project, had been part-funded by the billionaire George Soros's Open Society, and was usually "critical of governments that have fallen into disfavor with the United States and then are singled out for accusations of corruption" (Parry 2015f). Parry was suspicious of what he considered Bellingcat's collaboration with the: (1) Atlantic Council, a conduit for pro-NATO propaganda which partnered with Higgins for a report about Russian involvement in Ukraine; and (2) Soros-financed HRW (which, like Bellingcat, pushed claims of Assad government responsibility for the sarin gas attack in Ghouta in 2013). Parry claimed:

> Higgins has spread misinformation on the Internet, including discredited claims implicating the Syrian government in the sarin attack in 2013 and directing an Australian TV news crew to what appeared to be the wrong location for a video of a BUK anti-aircraft battery as it supposedly made its getaway to Russia after the shoot-down of Malaysia Airlines Flight 17 in 2014.
>
> (Parry 2015f)

In a later article, Parry opined:

> A central flaw in the Internet-based approach is that it is very easy for a skilled propagandist in a government dirty-tricks office or just some clever jerk with Photoshop software to manufacture realistic-looking images or documents and palm them off either directly to gullible people or through propaganda fronts that appear as non-governmental entities but are really bought-and-paid-for conduits of disinformation.
>
> (Parry 2015j)

Higgins gained mainstream acclaim, in part, because "his findings" always matched up with the propaganda themes peddled by the US government and its Western allies. Though most genuinely independent bloggers are ignored by the mainstream media, Higgins has found his work touted.

The Higgins version of the MH17 tragedy was partly borne out but also partly contradicted by a Reuters report in the *Moscow Times* (then owned by a European media corporation known for its often anti-Kremlin stance) in March (Reuters 2015b). This cited accounts from four villagers of Chervonyi Zhovten in rebel-held territory of Eastern Ukraine who said they saw a missile overhead just before MH17 was attacked, which seemed to come from separatist-held territory. The nearest Ukraine-held territory was 3.7 miles away. Chervonyi Zhovten lay seven miles north of Snizhne where Bellingcat.com and others claimed a Russian-made BUK had been positioned – which, if true, would suggest two BUK missiles in close proximity. One of the villagers at Chervonyi Zhovten claimed that a missile battery had been positioned in a near-by field on the day of the attack. This was corroborated by an ex-rebel (who had not actually witnessed the launcher) who claimed that rebel sources had targeted Ukrainian jets shortly before MH17 came down. But one witness later changed his story off-camera. The field from which the missile had allegedly been fired showed no evidence of a launcher. Rebels mostly claimed that they had access only to PZRK shoulder-launched missile systems, lacking the range needed to account for MH17.

## Australian *60 Minutes* investigation

The Higgins version was prominent in an Australian television *60 Minutes* investigative report in May 2015. *60 Minutes* followed what Higgins claimed was the trail of a camouflaged Russian BUK surface-to-air missile launcher from the base of Russia's 53rd Air Defense Brigade in the Russian city of Kursk which, one month before the attack, traveled into Eastern Ukraine and after the attack drove back to Russia with one missile missing. But why would the Russians drive the launcher back to Russia without the same camouflage they had used for its arrival (Vlemmix 2015)? The image of the so-called "getaway" was first provided by the Ukrainian interior ministry. Note that German intelligence, BND, had said in October that some images provided by the Ukrainians had been "manipulated."

*60 Minutes*' language was propagandistic. The presenter presumed he was investigating a "war crime" (not a mistake); he called Eastern Ukraine "Russian controlled" – contestable – and referred to "separatists and their Russian masters." The team visited a few months' ahead of publication of the official Dutch report, which it complained was moving "at a snail's pace." They did not ask relevant US sources why the USA had not released its own satellite evidence (Maybury 2015). Without evidence, they attributed responsibility unequivocally to Vladimir Putin. They did not ask why the Russians apparently failed to carefully paint over the entirety of the vehicle's three-digit identification number – instead, leaving two digits exposed – on their way to Ukraine, and then painted out more of the ID, though incompletely, in Eastern Ukraine. Images in different photographs of the rubber side skirt suggested similarity, certainly, but did not prove representation of the same physical reality as Higgins claimed. The "evidence" presented to viewers was not based on the actual, fuzzy, images but on a thick colored line painted over the images, which could show only what the artist wished to suggest. There were shots of what were asserted to be track marks on the road left by the launcher (nine months after the attack). Similar marks might have been left by other kinds of vehicle or even by the same vehicle at a different time. Burn marks were not found in the field from which the alleged missile was fired, but *60 Minutes* asserted they had been removed by ploughing and the passage of time.

A critical issue was not the purported route of the BUK, put together from images allegedly uploaded to publicly available Russian social media (which few non-Russians if any would be able to access and verify), but the determination of the trajectory of fire from the alleged launch point to the point of impact – MH17. While the Higgins story and *60 Minutes* provided an immediate and dramatic appearance of veracity, there were many opportunities in the methodology for the faking of images and bait-and-switch manipulation. The Dutch documentarist, Peter Vlemmix, found that much of the so-called social media evidence on YouTube had disappeared and that some of the images were actually Ukrainian images of their own Buk missile launchers, of which they had twenty-seven. He noted that MH17 was actually the third plane shot down in the same week. One was a cargo plane, leading many later planes to avoid flying over the area. So he asks, what led MH17 to change course over Poland so that it would fly directly over the war zone – was it Kiev Air Control? Parry (2015d) challenged the *60 Minutes* investigation, claiming that the image of the BUK missile returning to Russia was problematic. On some videos of lesser significance, he observed, the *60 Minutes* team did successfully superimpose its own shots over shots taken from social media to demonstrate similarity. But *60 Minutes* did NOT do that for the crucial "getaway" video. Instead, the *60 Minutes* correspondent, Michael Usher, simply stood in front of an intersection near Luhansk and claimed it was the same intersection that appeared in the "getaway" video, even though there were in fact significant differences between the two, a problem that Parry concludes was not rectified by a *60 Minutes* update.

The Higgins, Correct!V and *60 Minutes* versions were similar to a book-length account offered by the ex-editor of (pro-Ukraine) *Kyiv Post* and later UK resident, Jaroslav Koshiw. Koshiw believed that Donetsk militia removed the pilots' bodies for secret autopsies, destroyed the plane's cockpit and took the flight recorder, holding it for four days before handing it to Dutch investigative authorities (the final Dutch report noted no evidence of tampering). He concluded that MH17 was probably mistaken for an AN-26 transport plane. He attached supreme blame for this "war crime" (having previously called it a "mistake") to Vladimir Putin, but doubted the Dutch would pursue Putin given the sensitivity of relations between Russia and Royal Dutch Shell.

## First anniversary propaganda

MH17 propaganda revved up as the first anniversary of the tragedy approached in summer 2015. The Russian defense firm Almaz-Antey claimed that MH17 was hit by a BUK missile whose production they discontinued in 1999 but was now in Ukrainian hands. Further, the missile had been fired from Zaroshchenske, which Russia argued was then under the Ukrainian control. A story by the *New York Times*' David Herszenhorn on June 28 (Herszenhorn 2015b) alleged the reluctance of Luhansk separatists to assist investigators from Holland's Public Prosecution Office as they researched cellular towers and networks. Buried deep was an admission that separatists had helped the team achieve their main objectives. An anonymous Western source criticized the Dutch attempt to undertake such research under conditions of warfare. Unreferenced were likely local concerns for communications security in the event of opening up networks to foreign investigators.

Amidst claims and counter-claims was a lawsuit from a group of MH17 crash victim families against Igor Strelkov, commander of ethnic Russian forces in the area from which the plane was said to have been fired on. His was the voice heard celebrating, it was claimed, the shooting-down of a transport plane shortly after the MH17 crash, although the authenticity of available evidence was disputed. Robert Parry was astonished that US intelligence had issued no refinements of its understanding of the tragedy since July 22, 2014, speculating that official US silence allowed Putin-did-it propaganda campaigns free rein in WMM. Parry pointed to an entirely different though anonymously sourced narrative:

> A source who has been briefed on the outlines of the investigation said some US intelligence analysts have reached a contrary conclusion and place the blame on "rogue" elements of the Ukrainian government operating out of a circle of hard-liners around one of Ukraine's oligarchs. [...] I'm told that the reason for the DNI's reversal from openness to secrecy was that US intelligence analysts found no evidence that the Russian government had given the rebels sophisticated anti-aircraft missiles capable of downing an aircraft at 33,000 feet, the altitude of MH-17, and that an examination of US satellite and electronic intelligence instead implicated extremists linked to Ukraine's US-backed regime, although not to Kiev's political leadership.
>
> (Parry 2015h)

Parry later noted that shortly after the attack, the DNI had US intelligence analysts brief a few select mainstream reporters, "but the analysts conveyed much less conviction than their superiors may have wished, indicating that there was still great uncertainty about who was responsible" (Parry 2016a). In a *Los Angeles Times* story cited by Parry, "U.S. officials said it was possible the SA-11 [the designation for a Russian-made anti-aircraft Buk missile] was launched by a defector from the Ukrainian military who was trained to use similar missile systems" (Parry 2016a). One of Parry's sources briefed by US intelligence analysts – a "mainstream reporter"? – "claimed that as the investigation advanced they came to believe the attack was carried out by a rogue element of the Ukrainian military with ties to a hardline Ukrainian oligarch" (Parry 2016a).

Dutch intelligence, according to a report whose full details were not cited by Parry, believed as late as October 2015 that the only anti-aircraft weapons in Eastern Ukraine capable of bringing down MH17 at 33,000 feet belonged to the Ukrainian government. This might explain why commercial aircraft were allowed to continue to fly over Eastern Ukraine. The intelligence agency believed that the rebels had only short-range anti-aircraft missiles and a few inoperable BUK missiles that had been captured from a Ukrainian military base. Russia might have sent a BUK system into Ukraine or pro-Ukraine intelligence agencies exerted themselves to create evidence of a launcher that could be claimed to have traveled from Russia to a launch site in Ukraine and then traveled back again. Among the mooted reasons for why Ukrainians might have shot down MH17 was that a "poorly trained and undisciplined Ukrainian squad mistook MH-17 for a Russian plane that had penetrated Ukrainian airspace or that the attack was willful provocation designed to be blamed on the Russians" (Parry 2016b).

The ex-CIA analyst, Philip Giraldi, also cited the belief of some agency analysts that the most likely candidate might have been then governor of the Dnipropetrovsk Oblast, Ihor Kolomoiski. Kolomoiski was known to employ Israeli mercenaries as advisers and had personally organized and paid for militias to fight separatists. He could have organized a false flag incident to pin on the Russians or their surrogates, and had the means to do so. A television clip of an interview with Kolomoiski, found on the blog of the Dutch author of "MH-17: The Cover-Up Deal [*MH:17: de Doofpotdeal*]" Joost Niemöller (2014) could be interpreted as a confession that he was in a charge of a rogue operation designed to attack Russia's presidential plane but which accidentally hit the Malaysian airliner. Equally, Giraldi speculated, the Kiev government, acting independently, had the resources and motive (Giraldi 2015).

A report leaked to a private LiveJournal account – LiveJournal accounts were a favorite destination for Russia's disinformation apparatus, the Internet Security Agency – from a group described as "old hand aviation security experts" argued that MH17 was downed by an Israeli Python air-to-air missile intended for MH17 and for blame to be pinned on Moscow. The report disputed the claim that missile fragments extracted from the debris were similar to those used in warheads of surface-to-air Buk missile systems. It claimed that such "double-t strike elements" weighed more than twice as much as a single damage fragment among those that pierced MH17's hull. There were four air-to-air missiles that fitted the description of what could have taken down MH17 namely: French Magis-2, Israeli Shafrir, American AIM-9 and Israeli Python. The first three were eliminated as likely for various reasons, including type of warhead or guidance system specifications (RT 2015f).

The first anniversary was an occasion to stiffen demands for legal accountability and punishment. Both civil and criminal investigations had several more months to run. Yet the presumption of such conversations as represented in

WMM (sometimes backed by claims or opinions of identified or anonymous sources close to the investigations) was either Russian guilt or the guilt of separatist fighters, commonly regarded in the West as Russian-backed. Venues for justice were canvassed: national courts, the International Criminal Court in The Hague or the European Court of Human Rights. When Russia urged the United Nations Security Council (UNSC) and the International Civil Aviation Organization (ICAO) to be more pro-active in holding those responsible accountable, the *New York Times* Editorial Board retorted that Russia was trying to thwart the ongoing Dutch investigation (from which Russian experts had virtually been removed – in the view of the permanent representative of Russia to the UN, Vitaly Churkin, speaking on the Russian television channel, Rossiya 24 [UNIAN 2015c]), whose members, the Editorial Board predicted, would allege the culpability of ethnic Russians or Russian soldiers (*New York Times* Editorial Board 2015). The Russian foreign minister, Sergey Lavrov, complained that the Netherlands was withholding vital information from Russia, failing to explain why the Dutch security service made the announcement about the elements of the Buk missile only in August when they'd found them several months previously, and refusing to reveal fragments (Litvinova 2015).

Yet Russia vetoed a UNSC resolution (passed 11–3) of July 2015 for setting up an MH17 international criminal tribunal to prosecute those responsible. The resolution, coming from the USA and five US allies – Australia, Malaysia, Netherlands, Ukraine and Belgium – seemed calculated to escalate the US-led pressure on Moscow. Spokespersons had publicly reaffirmed Russia's opposition to the creation of a criminal tribunal, noting that no criminal tribunal had ever been established by the UN to deal with an air crash. Of the fifteen current members of the Security Council, eleven voted for the motion and Russia voted against, while three abstained – China, Angola and Venezuela. The resolution was more about punishment than investigation, and seemed to indicate – before the completion of two already ongoing investigations (neither of which had formally implicated Russia) – that the sponsoring parties had already determined guilt (Head 2015).

The *New York Times* derided – without evidence or reference to evident US-NATO presence in Ukraine – Russian claims of non-involvement in the Donbass (declaring that Russia "makes little effort" to conceal its "military support") and laughed off Russian concerns about Ukrainian fascism (in the very same month that Right Sektor fascists were threatening to bring down Poroshenko's government). The *New York Times* concluded that the consequence of all this was that Ukraine, so distracted by fighting in the east, was unable to repair its bankrupt economy or root out corruption (*New York Times* Editorial Board 2015).

Several sources looked for precedents from the case of the bombing of Pan Am 103 over Lockerbie, Scotland in 1988, thought to be the doing of Libyan terrorists. Libya had agreed to extradite two accused terrorists to an ad hoc tribunal established at an abandoned US military base at Camp Zeist in the Netherlands. These references rarely extended to well-documented doubts about the final judgment in this case (see Boyd-Barrett 2015). Another precedent, identified by Schrad (2015) was Iran Air Flight 655 from Tehran to Dubai, shot down in 1988 by an American surface-to-air missile from the USS Vincennes in the Persian Gulf, killing all 290 aboard. WMM references to this incident commonly presumed it was a tragic mistake. By contrast, MH17 discussions usually took the form of strident demands for criminal accountability. Iran had taken the USA before the International Court of Justice in The Hague. When American culpability for the tragedy seemed increasingly likely, the USA reached a settlement in 1996, paying $61.8 million to the victims' families. The USA never admitted legal liability, and no American serviceman was ever charged with a crime or served time.

## Dutch Safety Board report, October 2015

The Dutch Safety Board final report concluded that MH17 was most likely struck by a Russian-made surface-to-air BUK missile 9M38-Series with a 9N314M warhead, but the report sought to avoid "blame and culpability." The Dutch Safety Board (DSB), inexplicably, was unable to cite satellite evidence from Ukraine, Russia or the USA. A few months following publication the father of the only American citizen on board MH17 – 19-year-old Quinn Schansman – appealed to Secretary of State John Kerry to:

> hand over to the DSB radar and satellite data of the minutes before and after the crash. [...] This would enable the DSB to reopen the investigation and include a chapter with this information, which is essential for a successful criminal prosecution.
>
> (Cited by Parry 2016b)

The absence of the Dutch Safety Board's intention to assert culpability or direct evidence of such did not prevent WMM from strongly implying, if not asserting, the guilt of either Russia or "Russian-backed" separatists. Western and Ukrainian media frequently drew on reports from the controversial Bellingcat.com as justification or simply assumed that "Russian-made" was the equivalent of Russian responsibility – ignoring the extent to which Ukrainian weaponry was still heavily dependent on Russian sourced manufacture. The board was critical of the Ukrainian authorities for failing to close the airspace above the conflict zone. It concluded that the missile originated in an area of 320 square kilometers (123 square miles) in Eastern Ukraine. In efforts to locate the point of origin in a zone supposedly controlled by their opponents, Russian experts insisted that the area must be smaller, while Ukrainian experts claimed it was smaller still. This indicated considerable margin for doubt in the minds of Dutch Safety Board members.

On the day the DSB released its final report, the Russian Foreign Ministry responded that the "report is flawed, as the Russian findings were not taken into account" (Grozev 2015). The Deputy Chief of the Russian aviation agency Rosavia, Oleg Storchevoi, told journalists "the number of inconsistencies in this report is off the charts" (Hartog 2015). Storchevoi claimed that the shrapnel referred to in the Dutch report (likely collected by the SBU, a questionable source) was not identical to that found at the wreckage site. A press conference by Russian arms manufacturer Almaz-Antey attributed blame to a Buk missile that they had manufactured, while claiming that it was an older model than that indicated in the Dutch Safety Board report. A Kremlin spokesman, Dmitry Peskov, complained that Moscow had "repeatedly expressed its disappointment over the lack of cooperation and involvement of Russian experts in the investigation" (TASS news agency, cited in Hartog 2015).

Oleg Storchevoi wrote a column for Dutch daily *De Volkskrant* questioning the handling of the MH17 investigation and what he called the low quality of its "vague" final report. More specifically, Storchevoi charged that the DSB: (1) studied photos on social media when they should have been attempting to reach the site of the crash; (2) once they did arrive, they did not conduct field experiments in order to determine the exact sequence of events; (3) the Board refused offers of cooperation from Almaz-Antey, lodging only a single enquiry with the arms manufacturer, and did not visit Russia, despite numerous offers; (4) shifted the blame for Kiev's failure to secure airspace above the conflict zone to international carriers and regulatory bodies while (5) the Board chairman made public statements implicating Russians or separatists, going beyond the report itself and his competency as Chairman (RT 2016a).

Storchevoi's points (1) and (2) made little allowance for the genuine safety concerns that had prevented the investigating team from starting work at the crash site until late July, and required them to withdraw on August 6. Later investigation was plagued by impediments of access to the site, posed either by the separatists or the civil war. It was the separatists who recovered the flight data recorder and cockpit voice recorder and handed these fairly promptly, without tampering, to Malaysian officials in Donetsk. Not until August 2015 were possible BUK missile parts identified at the crash site. But access problems, however severe, could not compensate for the problematic assertions on the basis of unsound investigation.

In a critique of Russian protestations, Christo Grozev (2015) argued that Russia was an active contributor to the Dutch investigation, sometimes more successful in influencing the content of the final report than other parties, while withholding vital information. While Russia may have been an "active contributor," as Grozev argues, the team of investigators was led by the Netherlands and included representatives from six other countries – Ukraine, Malaysia, UK, USA, Australia and the Russian Federation.

Grozev's critique appeared on the Bellingcat.com website. He had managed the radio operations of a public US company in CEE and Russia, owned and operated national radio stations in the Netherlands and Ukraine, and was a senior researcher at Risk Management Lab, a think-tank at New Bulgarian University (Sofia) where he studied Russia-related security threats and the weaponization of information.

The Russian defense manufacturer, Almaz-Antey was invited by the Dutch Safety Board (DSB) to give evidence, evidence which sometimes contradicted the conclusions of other parties, including Bellingcat whose founder, Eliot Higgins, was dismissive of the quality of Almaz-Antey's findings with respect to the type of missile, the angle of impact and the location of the launch site. Russia participated in all three interim-progress meetings held by the DSB and provided seventy-four written comments between the draft and the final report. Nineteen of these resulted in textual amendments with varying degrees of significance.

Grozev's critique was suggestive of over-defensive Russian nit-picking. Alternatively, it could be concluded that the draft report in many respects betrayed a pro-Western framework of understanding, and that Russia had more work to do than other parties in trying to achieve a fairer accounting. Since WMM had framed Russia as the most likely culprit, Russia was correctly motivated to scrutinize all evidence meticulously.

Russia may have stepped beyond meticulousness in seeking the elimination of references that were not favorable to its position – for example, with respect to the evidence of the missile launch site. Russia seemed determined to show that this was not in separatist-controlled areas, an assessment that was not supported by all other sources (although some sources maintain that combatant lines were very fluid on July 17). These included the Russian opposition newspaper *Novaya Gazeta*, one of several media that had interviewed villagers of Zaroshchenske, the location preferred by Russia as the launch site and where it claimed the Ukrainians had positioned two BUK units on the day in question. Villagers asserted that there had been no Ukrainian forces in that area at the time.

Plausibly, the draft report had listed favorably towards partisan or controversial parties, including Ukraine and Bellingcat – a private source whose choice of investigations and their results neatly coincided with Western foreign policy interests. Grozev is too severe in criticizing Russia for changing its position on certain key issues, such as whether the missile came from the ground or from the air. New evidence and new thinking often leads to a change of position. Nonetheless, in returning to an air-to-air claim, Russia possibly stalled in providing the necessary supporting evidence. Which is not to say that the theory could not be correct.

The fundamental finding of the Dutch Safety Board's draft and final reports that the offending missile was fired by a Russian-make BUK, was challenged by parties to the Joint International Criminal Investigation Team (JIT), still ongoing, as represented to an inquest into the deaths of Australian victims in Victoria, in December 2015. Members of the Australian Federal Police and Dutch police and prosecutors said that DSB had failed to provide "conclusive evidence" of what type of munitions destroyed the aircraft. Their doubts linked back to statements by the Dutch State Prosecutor, Fred Westerbeke, that no more than twenty-five pieces of metal had been recovered, and by Australian pathologist David Ranson, in his reports to the Coroner, that no shrapnel had been identified in scans of the bodies of the crash victims and that this finding was inconsistent with the possibility that the plane had been hit by a BUK missile.

> Testifying for the first time in an international court, Detective Superintendent Andrew Donoghoe, the senior Australian policeman in the international

> MH17 investigation, said a "tougher standard than the DSB report" is required before the criminal investigation can identify the weapon which brought the aircraft down, or pinpoint the perpetrators. Their criminal investigation will continue into 2016, Donoghoe told the Victorian Coroners Court on Tuesday morning [December 15]. He and other international investigators are unconvinced by reports from the US and Ukrainian governments, and by the DSB, of a Buk missile firing.
>
> "Dutch prosecutors require conclusive evidence on other types of missile," Donoghoe said, intimating that "initial information that the aircraft was shot down by a [Buk] surface to air missile" did not meet the Australian or international standards of evidence.
>
> (Helmer 2015b)

Donoghoe told the inquest that the Australian Federal Police were unable to provide their forensic evidence because to do so would require Ukraine's permission – another party to the JIT – which had not consented to the disclosure of findings. In his final judgment, the Coroner, Iain West, did not refer to Donoghoe's written report nor to his courtroom testimony but, instead, cited unsourced forensic evidence in support of the DSB's findings that the culprit was a BUK missile, cause of 800 hits on MH17 by high-energy objects from a 9N314M warhead. One barrister charged that by excluding the evidence of the Australian Federal Police and of his own pathologist, "West has taken sides in the politics of the case. If on top of that he is refusing to release the Donoghoe and Ranson reports, he is covering up. West has discredited his own court" (Helmer 2015c).

The JIT's findings, as represented to the Australian inquest, would not likely have phased a senior officer of the Security Service of Ukraine (SBU) who in October told WMM that such missile fragments had been found in the bodies of victims (Sandhu 2015) – an assertion with which Australian pathologist David Ranson clearly disagreed. The SBU assertion commanded far greater attention in WMM than the Australian Federal Police's doubts and was likely intended to deflect public attention from the looming clash of view between Dutch Safety Board and the Australian Federal Police. Similarly, Australian media covered the Australian Coroner's reports with a lack of curiosity as to its inconsistencies:

> The corporate press in Australia has reported blandly Deputy Coroner Iain West's report on December 16, 2015 into the deaths of Australian victims of the crash of Malaysian Airlines Flight 17, including accepting at face value all the absences and contradictory statements by the coroner in his report. Below is an example of the mainstream reporting, in the form of the article which appeared in the *Sydney Morning Herald* (full text). The report by the Australian Broadcasting Corporation is equally bland and uninforming, though much better grammatically than that of its partner in disinformation.
>
> (Helmer 2015c)

WMM over several months developed the impression that the sequel to the DSB report would be a criminal investigation and this would blame Russia. Not only did alternative scenarios persist but some European politicians indicated that all might not be well with the criminal investigation. The problem was not only that much of the evidence was examined too late, but that it was examined by dishonest persons. At issue was the credibility of a deeply interested partner – Ukraine – a member of the Dutch Safety Board and the Joint International Criminal Investigation Team, a lot of whose evidence had been handled by the SBU, an institution many of whose members had been linked to crime. The SBU's former head, Valentyn Nalyvaichenko, resigned in June 2015 having been implicated in a stolen art scandal; over the following summer, twenty-two SBU agents were imprisoned for corruption and criminal practices (Van der Graaf 2015).

These considerations return us, then, to problems identified at this chapter's start, with initial claims that the separatists had Buk missile launchers and missiles. The US White House had asserted within days of the crash that it had satellite evidence to clinch such claims yet had never released the evidence, not even to the Dutch Safety Board. This suggested that the evidence in US hands actually contradicted the White House. No evidence of the presence of Buk missile launchers was referenced by the office of the Director of National Intelligence in its evaluation of the weaponry available to the separatists at that time. From the DSB report – Parry has observed and to whose account I revert – we know that Ukrainians insisted up until the shoot-down that rebels had no surface-to-air missiles that could threaten commercial airliners at 33,000 feet (a contestable claim, as Buk missile launchers were observed in separatist hands before July 17). Ukraine kept Ukrainian skies open to commercial air traffic, presumably on the basis of that intelligence. Yet the Ukrainians themselves had Buk systems that they were positioning toward the east. One day before MH17 was hit, a Ukrainian Su-25 fighter jet was shot down by what DSB reports show, on the basis of Ukrainian claims, was an air-to-air missile. Yet the skies remained open. Ukrainian air defense units would likely have been alert to the possibility of Russian jets entering or leaving Ukrainian airspace. Yet the DSB report does not ask why Ukrainians had turned on the radar they used to guide Buk missiles in the days before MH17 was shot down:

> A more likely scenario of the MH-17 shoot-down was that Ukraine moved several of its Buk batteries to the frontlines, possibly fearing a Russian airstrike, and the operators were on edge after a Ukrainian warplane was shot down along the border on July 16, 2014, by an air-to-air missile presumably fired by a Russian plane.
>
> (Parry 2015l)

Bellingcat released a report purporting to link the Buk launch vehicle that it had previously claimed to be present in separatist controlled territory of Eastern Ukraine to the 2nd battalion of the 53rd Russian Air Defense Brigade. The claims were made on the basis of the names and photos allegedly posted by soldiers having something to do with the convoy, and army data available online about personnel. It claimed that 100 soldiers were implicated, twenty of them directly, including the brigade commander and the battalion commander.

In this chapter, I have outlined the main claims and counter-claims concerning the possibility of Russian or ethnic Russian separatist responsibility for the shooting down of MH17 on July 17, 2014. I have sought to demonstrate that the most "authoritative" evidence, in the form of the Dutch Safety Board's preliminary and final reports did not attempt to assign culpability but did offer explanations for the crash that seemed to support Ukrainian and Western claims of how it occurred. I have suggested that the MH17 case, precisely because so tragic, has been deployed as a red-herring to distract attention from the main issues concerning the Ukraine crisis of 2014–2015 and, instead, beating up a lather of Western public repugnance for Russia in general, Vladimir Putin in particular, and the ethnic Russian separatists of Eastern Ukraine. But I have also cited many sources of doubt as to the real culpability for the shooting down of MH17, doubts which I do not expect to evaporate on the release of the final report of the Dutch-led Joint Investigation Team that was expected to report its findings of criminal responsibility in the course of 2016. Finally, I have emphasized a critical dimension, one that is articulated in the DSB final report but has been significantly down-played in WMM namely: the failure of the Ukraine authorities to close down airspace to civilian airliners over Eastern Ukraine at a time when it was apparent that the war zone constituted grave danger to all aircraft, regardless of altitude. The concept of criminal negligence barely seems to cover it.

# 9 The Russians are (maybe/probably not) coming![1]

There was no time in the conflict when anyone would have been overwhelmed by the thought that Russia might be supporting ethnic Russian separatists in the Donbass regions of Donetsk and Luhansk, now self-declared independent republics. Russia was immediately adjacent to these regions and shared a long border with Ukraine. Russia and Eastern Ukraine shared strong cultural affinity. The ethnic Russians of Eastern Ukraine had reason to dread a regime catapulted into existence by anti-Russian forces of Western Ukraine, a regime with little political support in the east, and that indulged aggressive rhetoric against ethnic Russians, and the Russian language and culture while glorifying an extreme nationalist perspective on Ukrainian culture. In the midst of fighting between Ukraine and the separatists, Ukraine turned its back on the independent republics, ceasing all financial support, including pensions and social security payments. There was still a measure of sympathy in the east, especially among older generations, for the culture and security of the former Soviet Union. The people of Eastern Ukraine at many points in this conflict were desperate for food, shelter and protection from aggressive Ukrainian shelling of civilian populations. So it was not so much a case of wondering whether Russia would invade eastern Donbass as a question of why on earth would it not.

The reality, however, was more complicated. Despite popular support for Russian annexation in Crimea, this was less the case in eastern Donbass where many preferred greater autonomy within an over-centralized Ukraine to annexation by Russia. Knowing this, and not wanting to exacerbate an international crisis with a measure that would be less defensible in international law than the case of Crimea, Russia proceeded delicately on the eastern Donbass question. It resisted talk of annexation; it tried to discourage the voting that led to the people's republics and later sought to postpone local elections there until after completion of the process of constitutional reform in Ukraine that had been demanded by the Minsk II agreement.

The West and WMM stoked up fears of a full-scale Russian invasion of eastern Donbass that never took place, and NATO leaders persisted in such talk well into 2015, even to the point of alienating key European allies who grew tired of evidence-free fear-mongering. In the absence of an invasion, Ukraine and NATO searched desperately for evidence of lower levels of Russian involvement, from the presence of Russian military and volunteers, to Russian military equipment and other forms of assistance. As inevitable as these might have been, the actual evidence was generally less impressive than the claims.

## Humanitarian convoys

The situation by mid-August 2014 had severely deteriorated for ethnic Russians and separatists in the Donbass region. Following Ukrainian army bombardments, 250,000 were without water, electricity or the telephone (Kramer 2014b). Faced with a dire humanitarian catastrophe (which excited next to no sympathy in Western powers), Russia readied a 260-truck humanitarian convoy. This was halted at the border for several days while both sides, Ukrainians and Russians, negotiated the terms under which the convoy could proceed. By August 19, the Ukrainians said the convoy could proceed but would not guarantee its safety in areas not under its control.

Journalists working for the *Guardian* and *Telegraph*, supposedly stationed on the border, reported on August 15 that they were "eyewitnesses" as a convoy of military trucks crossed the Russian border into the breakaway Donetsk republic. While there had been photos of military trucks accompanying the humanitarian convoy while it was on Russian territory, there had been no proof so far – aside from the said eyewitness reports – that confirmed Russian military vehicles, with or without a humanitarian convoy, either entering or inside Ukraine. The following day, Ukraine military's spokesman, Andriy Lysenko, reported that Ukraine forces had attacked an armed convoy from Russia, and "destroyed" a part of it. This was subsequently reiterated by both the president of Ukraine and NATO's secretary-general who claimed that the alliance had detected an "incursion" of vehicles from Russia, adding that "what we have seen last night is the continuation of what we have seen for some time" (Walker 2014b). Missing from WMM discourse was critical concern for definitions. What was an "armed convoy?" Could the term include a humanitarian convoy under the protection of a few armored vehicles? Even less in evidence was concern for potential loss of Russian lives.

There was zero actual evidence of the attack on the Russian convoy. Russia denied the claims. Durden (2014a) wondered whether Ukraine had accidentally destroyed some of its own vehicles. The leaders of the Donetsk People's Republic (DPR) dismissed Ukraine's claim to have destroyed a convoy of Russian armored vehicles: "We haven't received any armored vehicles from Russia. No Russian units, including Russian armored vehicles, have crossed the border. Hence, no Russian armored vehicles could have been destroyed," DPR First Deputy Prime Minister, Andrei Purgin, told Interfax (2014). Purgin claimed, on the contrary, that ethnic Russian militias had destroyed about 100 Ukrainian armored vehicles. The White House admitted that while accusing Moscow of "incursions," it could not confirm that the convoy was destroyed by Kiev.

Whether British journalists who sourced the story of a Russian invasion lied or had been duped by the CIA (as suggested by Roberts 2014e), other media followed along without checking:

> The disreputable BBC hyped the story without investigating. The German media, including Die Welt, blared the story throughout Germany without concern at the absence of any evidence. Reuters news agency, also with no investigation, spread the story. Readers tell me that CNN has been broadcasting the fake story 24/7. Although I cannot stand to watch it, I suspect Fox "news" has also been riding this lame horse hard. Readers tell me that my former newspaper, The Wall Street Journal, which has fallen so low as to be unreadable, also spread the false story. I hope they are wrong. One hates to see the complete despoliation of one's former habitat.
>
> (Roberts 2014e)

Why might the story be false? Roberts summarized the arguments. The Russian government was clearly attempting to de-escalate the crisis. Putin had refused to support the claims of ethnic Russian separatists for unification with Russia. He had asked the Russian Duma to rescind his authority to intervene militarily in Ukraine. Russian military forces would not be sent into Ukraine prior to the Duma renewing Putin's authority so to do. Even if Russia had decided to invade Ukraine, it would not have sent one small armored group unprotected by air cover but a force capable of quickly rolling up Ukrainian forces. Not a single Western news organization hyping the story, Roberts said, had presented evidence to support the claims of a Russian invasion.

The White House detected Russian military vehicles painted to look like civilian trucks: even if there was evidence for this, and there wasn't, it would have been compatible with the possibility of reusing military vehicles for aid purposes. Western and Kiev's irritation with the humanitarian convey was ascribed to Russia's allegedly unilateral decision to cross the border without Red Cross escorts. The involvement of Red Cross escorts had apparently been agreed in principle. Their eventual exclusion was used by the Ukraine and NATO to support their description of Russia's action as a violation of Ukrainian sovereignty. This condemnation made no allowance for the long delays to which the convoy had been subjected by Ukraine and the urgency with which aid was needed. While the International Red Cross had "taken all necessary administrative and preparatory steps for the passage of the Russian convoy," it sought assurance of safe passage that was not forthcoming from Kiev (McGovern 2014b). Kiev refused a ceasefire or political settlement until the separatists had surrendered – impossible in the circumstances – but instead intensified its bombing of Luhansk (the convoy's destination). In Luhansk, residents had been without electricity, running water and a telephone service for nearly two weeks. People risked being torn to bits by mortar fire as they waited in long lines for food and water. In Donetsk, Ukraine was reported to have used white phosphorus (Williamson 2014), a chemical weapon. Russians and Ukrainians disputed whether and to what extent the trucks had been inspected beforehand. The convoy appeared to return immediately after the delivery of provisions.

A *New York Times* account by Andrew Higgins and Michael Gordon on August 22 launched with the claim that Russia had:

> escalated tensions with Ukraine to the highest level since its stealthy invasion of Crimea in the spring, sending more than 200 trucks from a long-stalled aid convoy into rebel-held eastern Ukraine over the objections of Kiev and, NATO said, conducting military operations on Ukrainian territory.
>
> (Higgins and Gordon 2015)

Without qualification, the authors depicted the delivery of humanitarian aid as an "escalation" of tensions, without reference to why the aid was necessary, namely, the destruction of ethnic Russian neighborhoods and the dire threats to civilian populations from missile and other attacks by Ukrainian forces. The story equated aid delivery with a "military" operation, yet was unable to support that claim. By contrast, the ex-CIA analyst, Ray McGovern, characterized the convoy "more like an attempt to break a siege, a brutal method of warfare that indiscriminately targets all, including civilians, violating the principle of non-combatant immunity" (McGovern 2014b). The *New York Times*' misleading reference to Russia's earlier "stealth invasion of Crimea" was provocative. Prior to the coup in Kiev, Russia legally stationed large numbers of troops in Crimea to protect its Sebastopol fleet. It did not need to "invade." The annexation of Crimea followed a popular referendum called by the people.

Higgins and Gordon conflated the issue of the convoy with a separate allegation, attributed to senior NATO and anonymous "Western" sources, that Russia had moved artillery units into Ukraine. Russian denials were treated with skepticism and likened to the "unidentified gunmen who paved the way for Moscow's annexation of Crimea – until President Vladimir V. Putin stated in April that Russian troops were 'of course' involved" (Higgins and Gordon 2015). Thus the *New York Times* falsely attributed Crimea's annexation to mysterious "unidentified gunmen," and treated it as Russian-initiated rather than something urged on Russia by Crimea. The Higgins and Gordon reference to "unidentified gunmen" tried to vilify Putin, first by sneakily suggesting without evidence that these were Russian troops, and second, by deceptively suggesting that Putin's acknowledgement of the involvement of Russian troops was a confession of complicity, whereas Russian troops were legitimately present in Crimea before the coup took place in Kiev.

The problems of WMM reporting of the initial convoys do not refute the possibility that Russians may have abused their access to the Donbass. It is also possible that the pretext of humanitarian aid was mainly authentic earlier in the hostilities, but became less so over time. Convoys continued into the following year. Though attracting less WMM attention, Ukrainian sources continued to express anxiety about what was in the convoys, what they brought back (the bodies of Russian soldiers?), and the restrictions on, or practicality of, inspections. A report of the UN Human Rights Monitoring Mission in Ukraine in September 2015 worried that the Russian Federation had "continued to send white-truck convoys without the full consent or inspection of Ukraine, and their exact destination and content could not be verified" (United Nations Human

Rights 2015). By this time, it declared, the Russian Ministry for Emergency Situations had sent "at least 40 convoys of 60 or more cargo trucks each into rebel-held areas and offloaded their contents out of sight of the dozens of monitors deployed in the region by the OSCE" (Williams 2015b). Due weight needs be given to the consequences if, in view of the complexity of the Minsk II agreements and the time it was taking for their implementation, Russia had felt encouraged to think it had free rein in these territories. As to what constituted "aggression" and "defense" was far from clear.

## Border "build-ups" and crossings

Writing on August 23, McGovern (2014b) considered it likely that many Russian citizens had crossed the border in support of ethnic Russians, even though Moscow had denied dispatching or controlling them. A new factor was the presence of Russians who openly claimed to have been sent by Moscow into Ukraine. There were regular Western defense sources for claims of a massive Russian military build-up of troops on the border, echoing a propaganda campaign that had marked the early days of the crisis. Estimations of the number of Russian troops involved ranged from 10,000 to 25,000. Allegations in mid-August that twenty-three Russian armored personnel carriers had crossed only to be destroyed by Ukrainian artillery proved totally unfounded, even while WMM repeated the unsubstantiated charges as if factual (Parsons 2014a). Western sources unreasonably presumed that violent action by Ukrainian military forces against ethnic Russian Ukrainians, close to Russia's border, should present no problem for Russia.

In late July, however, OSCE had deployed sixteen observers to two border-crossings in Russia – Gukovo and Donetsk – following a request by the Russian government. OSCE reported nothing. The Russian troop build-up would hardly have been difficult to discern (Lisitsyn 2014); yet no such evidence emerged. (The following year, OSCE, according to Kiev reports, did castigate ethnic Russian separatists for attacks on OSCE monitors [UNIAN 2015e]).

The USA repeatedly complained about "Russian aggression" in late June but failed to present verifiable proof (Parry 2014p). Within days of the downing of MH17, the US State Department released satellite images showing what it purported to be self-propelled artillery of a kind only to be found in Russian military units on the Russian side of the border, oriented towards a Ukrainian military unit within Ukraine. The reporter, Derek Monroe, told TRNN that such artillery could also have come from Ukrainian army defectors to the rebel side, bringing with them Russian equipment previously supplied to Ukraine or sold on the black market (Monroe 2014). While one could understand why Russia would provide moral, even logistical support to ethnic Russians, it would not be in Russia's interest to encourage a significant destabilization of Ukraine. In July, Ditz (2014c) reported that the US government was inventing a narrative of massive Russian artillery strikes against Ukrainian military bases along the border. Parry (2014b) reported that according to his sources, Moscow provided some covert support for rebels (mostly light weapons), but that Putin favored a political settlement in which Kiev would grant Eastern Ukraine more autonomy and accept Russia's annexation of Crimea in exchange for peace in the east and some financial support from Russia for the Kiev government.

Reporters for the British *Daily Mail* on August 20 (Stewart *et al.* 2014) asserted that Ukrainian security officials denied that a huge Russian military convoy was deployed in the city of Luhansk as had earlier been claimed by Lt-Gen. Igor Voronchenko, head of the Ukrainian Anti-Terrorist Operation (ATO) in the city, and backed by military analyst, Dmitry Tymchuk. This claim was even refuted by Kiev's National Security and Defense Council spokesman, Andriy Lysenko, who dismissed it as "strange." The *Mail* noted that Kiev received "advice" from Western spin doctors and "PR specialists." A month after the Kiev coup, a group of Kiev-based PR professionals had formed the Ukraine Crisis Media Centre. The information war was further muddied by Russian military movements in Russian territories close to the border with Ukraine, hard-to-substantiate tweets and by separatist brags – that may or may not have been true – claiming receipt of either or both equipment and personnel aid from Russia.

While the White House and WMM peddled fantasies as to Russian attacks on Ukraine, they paid scant attention to Ukrainian attacks on territories controlled by ethnic Russians and which had claimed hundreds of lives. They downplayed the military assistance to Ukraine furnished from the USA, NATO, the EU and possibly Polish sources. Even less attention was given to the role of large bands of neo-Nazi paramilitaries deployed to follow up army bombardments with urban warfare, raising the question: if Russia was indeed seriously interested in defending ethnic Russians by providing military and other support, why had its aid been so ineffectual? Parry commented that:

> As the Ukrainian military tightens its noose around the remaining rebel strongholds, battering them with artillery fire and aerial bombardments, thousands of neo-Nazi militia members are again pressing to the front as fiercely motivated fighters determined to kill as many ethnic Russians as they can. It is a remarkable story but one that the mainstream US news media would prefer not to notice.
>
> (Parry 2014q)

Leaders of separatist governments and militaries in Donetsk and Luhansk reportedly resigned in mid-August in a possible bid to render the DPR-LPR leadership more acceptable to Ukraine and lay the groundwork for a cessation of hostilities. Previous leadership had included Russian nationalists with possible ties to the Kremlin and/or to right-wing movements in Moscow – the term "right-wing" in this context referring to either or both extreme nationalist sentiment or to communist orthodoxy – that were exerting pressure on Putin for more active intervention (Peters 2014).

More than 2,000 were reported dead in the conflict thus far, with thousands wounded. Slain soldiers and civilians were totaled by Kiev as approaching 6,000 in February 2015, although German intelligence considered the real number to be closer to 50,000 (RT 2015a). The UN cited 8,000 in a report of September 2015. Over half a million and perhaps a million had fled to Russia. There were appeals for humanitarian aid amidst rounds of meetings and talks between all parties involved. By late August 2014, it was indeed beginning to appear, according to WMM, that Russian forces had invaded the Ukraine, although the quality of sourcing fell a long way short of compelling first-hand evidence and was subject to numerous challenges. It is not impossible that during this period, August through early September, there was an increasing presence of Russian artillery and personnel in Eastern Ukraine. But where there was doubt, many WMM professed certainty.

## The "August Invasion"

Bearing in mind the overwhelming staged-for-media spectaculars of numerous US invasions of other sovereign countries (including Afghanistan 2001, Iraq in 1991 and Iraq again in 2003), the "Russian invasion" was a strangely under-sourced, drawn-out, subtle and barely discernible event, if it happened at all (I am inclined to think that nothing occurred worthy of the term "invasion"). While Kiev and Western partners protested all along up to August that Russia had "invaded" Crimea and was "behind" the unrest in the Donbass, it was not until August that apparently serious claims of invasion were heard, overlapping as we have seen above, with anxieties concerning the Russian aid convoys.

The Ukrainian security services claimed to have captured ten Russian paratroopers in August, identifying them as soldiers of the 331st regiment of 98th airborne division based in central Russia (AFP 2014). They were taken near the Ukrainian village of Dzerkalne, about fifty kilometers (30 miles) southeast of Donetsk (AFP 2014). Moscow claimed that the soldiers had strayed unintentionally into Ukrainian territory. Within days, Ukraine returned them to Russia. Later, NATO released satellite images which allegedly showed mobile Russian artillery units and supporting vehicles "engaged in military operations inside the sovereign territory of Ukraine" (Queally 2014). The images came from Digital Global, an independent US-based satellite imaging firm. Since both sides of the fighting (Ukrainian and separatist) used Russian equipment, the images could not prove the presence of Russian troops. A *Moscow Times* story on February 6 attempted to pin-point items of Russian equipment that probably could not have come from Ukrainian armories but which had been sighted in the separatist Donbass (Bodner 2015a). Some of these "sightings" were likely based on dubious social media evidence. The items included T-72 main battle tanks, BMP-2 infantry fighting vehicles, SA-11 Buk air defense systems, BM-21 grad missile launchers and 259 Nona self-propelled artillery (Bodner 2015a).

Following the announcement of a ceasefire on September 5, Amnesty International, in a report that castigated both Ukrainian and ethnic Russian fighters for war crimes, claimed satellite evidence (sourced, once again, to Digital Global – the evidence was possibly sourced to the same tranche of images that the US government had already used) of a build-up of Russian amour and artillery and evidence of "direct Russian interference," as well as support for ethnic Russian separatists (Shetty 2014). The images purportedly showed new artillery positions being established just inside the Ukrainian border between August 13 and 29, and a similar artillery position in a field northeast of the first, also within Ukrainian territory. Imagery from August 26, 2014 showed six relatively advanced self-propelled howitzers (likely 2S19 Msta-S 152-mm) in firing positions facing southwest at Ukrainian army locations. While none of this constituted proof that the weaponry was actually controlled or even provided by Russia or that there were Russians present, Amnesty International considered the evidence compelling when added to reports of Russian incursions (the ten paratroopers?) and eyewitness accounts from those "fleeing the fighting" (no detail provided as to whose "eyes" were looking). Reports, mainly sourced to Ukrainian and NATO sources in the closing weeks of August, claimed that Russia had sent 1,000 troops, tanks and artillery into Ukraine, and Western analysts debated whether Russia was seeking to build a land bridge along the Azov Sea between Russia and Crimea. Many reports of fighting around Mariupol on the Azov Sea also assumed a Russian presence. These provided support for and context to reports that NATO intended to assemble an expeditionary force comprising 4,000 to 10,000 troops from seven different member states.

Exasperated, Roberts protested this narrative of escalation:

> Ask yourselves this question: Is the entirety of the Western media so ignorant and incompetent not to realize that a Russian invasion of Ukraine would not consist of 1,000 troops (the evidence of which no one can find), or is the entirety of the Western media simply willingly serving as a propaganda ministry for Washington's warmongers, as the Western media did for George W. Bush's invasion of Iraq based on fabricated evidence of Iraqi weapons of mass destruction? There are only two conclusions that can be reached about the Western media: Either it is completely stupid or completely corrupt.
>
> (Roberts 2014d)

One apparently knowledgeable but anonymous source, a Florida-based Russian blogger, interviewed by Mike Whitney, a frequent contributor to progressive or left-wing alternative media sites, offered the assessment that Russian troops had not deployed across the border. Instead:

> Moscow did allow volunteers and weapons to flow in. And while the assistance was not provided directly by the FSB (Russia's Federal Security Service) or the military, it was provided by various private groups. Clearly, the Kremlin has the power to help-out when it chooses to do so. In one instance, there appears to have been direct artillery support from across the Russian border (in the so-called "southern cauldron"), but most of the aid has been covert. Besides the covert assistance, Russia has also provided

> intelligence, logistical and political support for the Novorussians. Without Russia's support, the Novorussians never would have been able to turn the tide in the war.
>
> (Whitney 2014h)

A delegation visiting the Washington office of Oklahoma Senator Jim Inhofe in November – comprising Ukrainian members of parliament, a paramilitary leader and one Georgetown University professor – provided the senator with photos purportedly showing the Russian military invading Ukraine in the previous August and September. These were quickly debunked as photos taken from other conflicts: for example, one was an AP photo from the Russia-Georgia war in 2008 (Kredo 2015).

## Bellingcat.com allegations

In the preceding main section of this book, I have discussed the controversial role of the Bellingcat.com investigative journalism site in the coverage of the shooting down of MH17 as well as in contributing evidence during the 2013 saga of the Assad regime's alleged use of chemical weapons against Syrian rebels – allegations, eventually determined unsound, that almost served as a pretext for direct US and European intervention at that time. In February 2015, this site came into focus once more, this time on the topic of firing that allegedly occurred during July and August 2014 from alleged Russian rocket-launchers positioned on the Russian side of the border but targeting Ukrainian troops, with hits that the site's authors deemed to have been "decisive" in the three battles they investigated. A *Guardian* story by Julian Borger, the newspaper's Diplomatic correspondent, and, curiously, co-authored by the founder of Bellingcat.com, Eliot Higgins, claimed to have "near conclusive proof" of firing from inside Russia (Borger and Higgins 2015).

The investigators worked from crater patterns in satellite photos, and traced trajectories back to likely firing positions where they found scorch marks and tire tracks consistent with Russian rocket-launchers. Their conclusions were buttressed from estimations made on the basis of a number of videos of firings conveniently uploaded to YouTube by local residents. The story included the assessment of an expert from Cranfield Forensics Institute (Stephen Johnson) who considered that the methodology of crater analysis was "highly experimental and prone to inaccuracy." Perhaps this would help account for the fact that such evidence was not forthcoming from NATO or Kiev. Johnson conceded that there were markings consistent with scorch marks and unlikely to have been caused by agricultural machinery. The methodology raises the question, in this author's mind, as to whether the authors were in a position to share information concerning the range of possible firing positions – as opposed to only the positions that most conveniently supported their thesis.

A few days following the Bellingcat.com report, an article in *Sputnik* (previously *RIA Novosti*, financed by the Russian government) (Hirst 2015) reported that the *Guardian* had subsequently added a footnote to its story acknowledging that the co-author was indeed the founder of Bellingcat.com, and also the lead author of the report on which the story was based. *Sputnik* further reported that Higgins was a research fellow with the UK-based security and intelligence organization CENTRIC, based at Sheffield Hallam University. CENTRIC's website, as cited by *Sputnik*, boasted that it enjoyed a "close collaboration in security research and activity started between Sheffield Hallam University and law enforcement agencies," and whose board comprised individuals from UK policing and the British intelligence community (*Sputnik News* 2015a). Higgins had denied that CENTRIC had any involvement in his report on the Russian rocket-launchers, and that he had any link with the National Endowment for Democracy (NED). The *Sputnik* report quoted ex-CIA agent Robert Baer to the effect that information of the kind reported on by Higgins would require human resources on the ground if it was to be considered reliable. *Sputnik* further noted that Bellingcat.com was owned by Higgins' company Brown Moses, established in December 2013. *Sputnik* referred to use of the Bellingcat.com report by the UK Foreign and Commonwealth Office whose own site included supporting photos but which included, *Sputnik* argued, data on time and place that did not correspond with information on the very same photos published on other sites. A further consideration in an assessment of Bellingcat.com is the selectivity of its investigations: where was Bellingcat, for instance, in helping determine the trajectory of missiles fired on Mariupol in late January 2015 (see below)?

## OSCE

OSCE had been appointed to monitor the border between Russia and Eastern Ukraine. OSCE comprises fifty-seven participating States that span the globe, encompassing three continents – North America, Europe and Asia – and more than a billion people. Towards the end of October, the organization's monitoring chief, Paul Picard, confirmed that since the beginning of its observer mission at the end of July, it had not recorded any movement of military equipment or units from Russia into Ukrainian territory. One observer noted that the OSCE assessment nullified Washington and NATO claims of Russian invasion and infiltration of Ukraine, yet this was given negligible reportage in the Western media (Cunningham 2014). Cunningham cited an RFE report that many young Russian men had been recruited through the Internet to fight voluntarily and at their own expense alongside ethnic Russians in Ukraine. He also debunked NATO claims of an escalation of Russian flight incursions across Latvia and other non-Russian territories which were likely based, not on the actual number of incursions but upon the number of NATO "scrambles."

Reviewing the question of Russian intervention over the period of March to August 2014, Sakwa found the extent of Russian materiel and personnel to be indeterminate, such evidence as there was being far from conclusive before August. Indeed, he considers that the scale of this was "greatly exaggerated" by Kiev and the West. What is certain, Sakwa claims, is that the Donbass uprising was primarily homegrown. Yes, there were volunteers of various stripes, actually representing a danger to Russia itself not dissimilar to the danger of volunteer militants in Syria whose activities could easily blow back onto their sponsors (Sakwa 2015: 154–155). Russia trained and filtered some genuine volunteers, he believes, as well as regular forces in the capacity of "advisors." Only in August, he claims (his sources are unclear) did Russian "volunteer" paratroopers apparently take part in regular battles. This is not far distant from the evidence supplied later from sources sympathetic to the Ukrainian side, including Bellingcat.com and another so-called open-source researcher, Askai707 (as reported by US propaganda site RFE/RL) who tracked down many soldiers of Russia's 6th Tank Brigade by following their cybertrail of social-media postings on VKontakte, the biggest Russian-language social-networking site. This was when ethnic Russian forces surrounded Ukrainian troops at Llovaysk in the Donetsk region. There were 366 Ukrainian soldiers killed, according to the Ukrainian military (Wesolowsky 2015). These kind of reports, however, becoming available so long after the events, are difficult if not impossible to confirm. In the meantime, a constant refrain from the resistance movement was its need for more supplies and support, indicating the insufficiency of any support of any kind that it did receive.

Ukraine's President Poroshenko announced a unilateral ceasefire of one week on June 20, followed by multilateral talks on June 23. On that day, the Russian Federation Council revoked the ruling of March 1 that had authorized Russia to deploy troops on Ukrainian territory. This measure may then have actually encouraged a wide Ukrainian offensive against the Donbass with the enthusiastic support of a "war party" in Washington that included the US ambassador at the UN Security Council, Samantha Power, who for this event suspended what Stephen Cohen referred to as her "revered 'Responsibility to Protect' doctrine" (cited by Sakwa 2015: 164). The attacks consisted primarily of long-range artillery bombardments against apartment blocks and villages. As a consequence, there were large numbers of civilian casualties, a flood of hundreds of thousands of refugees who streamed out of the region, some of them for Russia, and every appearance of humanitarian catastrophe.

Following talks and attempted talks throughout July and August, and just prior to the announcement of a ceasefire in September, Western media reported "Russian" mortar attacks on the Ukrainian city of Mariupol, although it was not clear whether the reference to "Russia" was to Russian soldiers, pro-Russian or ethnic Russian separatists or federalists, or simply to Russian armor and equipment (the basis for both Ukrainian and pro-Russian or separatist weaponry). More likely the attacks were shells fired by pro-Russian separatists at Shyrokyne, close to Mariupol. A ceasefire was announced following a meeting in the Belarus capital of Minsk on September 5 between Presidents Putin and Poroshenko, Ukrainian and ethnic Russian representatives. The terms of the ceasefire included cessation of fighting, withdrawal of troops from the recent battlefield zones, and moves towards greater autonomy for ethnic Russian regions, with respect for the Russian language. In May 2015, Mariupol appeared to be still under Ukrainian control, even though pro-Ukrainian activists were reportedly in a minority. Ukrainian sources complained that the city's authorities were lax in their support of measures designed to protect the city from actual and possible future attacks from pro-Russian separatists or from regular Russian troops. The same sources complained that OSCE was "promoting Russia's agenda [...] repeatedly failing to see Russian troops and equipment in Ukraine and by recording mostly Ukrainian responses to separatist shelling" (Sukhov 2015a). Ukrainian media such as UNIAN sought to reverse the balance by focusing almost exclusively on reports of ethnic Russian shelling of Ukrainian positions.

## From Minsk I to Minsk II

One factor that may have encouraged caution on both sides was the acknowledged danger (Noack 2014) of the presence in Ukraine (home of Chernobyl, location of one of the largest nuclear reactor crises in history, surpassed only by Fukushima in 2011, and threatened by forest fires the following April) of fifteen nuclear power stations. The closest of these to the fighting was the Zaporizhia nuclear power station with six separate reactors. There were worries about the safety mechanisms in place at these power plants. Sergej Boschko, head of Ukraine's nuclear regulatory agency, was cited as telling the German public broadcaster, ARD, that "no nuclear power plant is protected against military attacks. They are not made for war, they are made for peace" (Noack 2014).

Another consideration was the approach of winter, given that Ukraine was still dependent for approximately half of its energy needs on Russian sources. Perhaps the biggest concern was the state of the Ukrainian economy. Inflation was at 20 percent, and the hryvnia had plunged by 40 percent since early 2014. The breakaway Donbass region removed 16 percent of GDP; Crimea represented the loss of a further 1 percent. Industrial production was down by 85 percent in Luhansk, and by 60 percent in Donetsk. Luhansk and Donetsk account for 95 percent of Ukraine's coal produce and for a high percentage of its exports (C.W. 2014). In deep depression, any forthcoming IMF funds, already subject to structural adjustment conditions bound to impoverish the majority of the population, would serve to pay only outstanding bills and feed the military machine whose fighting was taking place on Ukraine's industrial heartland (Escobar 2014c).

The US-imposed pressure on the EU to activate sanctions on Russia was also hurting European economies. German growth had sunk back to 1.5 percent while other European countries were returning to recession status. According to an industry lobby group, German exports to Russia slumped by 34 percent in the first five months of 2015. Before sanctions, there had been 6,200 German firms with an active presence in Russia. There were 300,000 German jobs linked to exports to Russia (Reuters 2015f). The US-NATO inspired "Ukrainian crisis" had engendered divisions within the EU, not just over the February coup in Kiev but over continuing US-NATO rhetoric about the "Russian threat," and the need to expand NATO by encircling Russia with NATO bases in East European countries. France, Italy and Spain were trying to resist this pressure; Germany was ambivalent, while the UK, Poland and the Baltic states – Estonia, Latvia and Lithuania – were strongly in favor. Meanwhile some observers, including the former British ambassador and Oxford academic William Brenton, considered that Putin was using Western sanctions to strengthen his domestic position and negate the political fallout from low oil prices by blaming the failing Russian economy entirely on the West (Amos 2015).

The ceasefire did not hold well. In early October, the *Los Angeles Times* reported that pro-Russian separatists and Ukrainian government forces had abandoned the ceasefire and separatists had stepped up their assault on Donetsk airport (Williams 2014). Ukraine rained bombs on Donetsk, one of which was possibly responsible for the killings of twelve civilian bus passengers on their way to Donetsk on January 13, 2015. Missiles fired on Mariupol – according to the unanimous determination of the City Government and numerous local testimonies – appeared to have been fired on ethnic Russian areas, following a flight path over the town of Sartana from Ukrainian held positions in the north (Eliason 2015a). Utilities were cut off a few hours in advance of the attacks. A new ceasefire had been agreed in December but appeared to have disintegrated by mid-January 2015 when the four-way talks in Berlin between Germany, France, Ukraine and Russia stalled. Conversations continued, with a meeting between German and French premiers and the Russian foreign minister in the first week of February.

Washington, meanwhile, noisily threatened to provide direct military aid and "training" to Kiev while Russia responded that any such move would require it to intervene militarily in order to secure its interests, thus enhancing the likelihood, even the inevitability of a world war. Even now, with Eastern Ukrainian separatists making steady military advances, surrounding 9,000 Ukrainian conscripts around Debaltseve (and amidst a clamor for war from the USA and some sections of the EU), there was significant room for doubt as to whether there was an actual Russian presence in Eastern Ukraine – reports in mid-February suggested that Kiev itself was increasingly ambivalent. One commentator noted (Dejevsky 2015) that this was a region where many Soviet-era armaments factories were sited, suggesting an abundant supply of "Russian" weapons in the region already without Russia necessarily being called on or actually supplying more. Even if Russia did have regular troops in Eastern Ukraine, the numbers were not significant enough to provide the raw evidence in the form of definitive satellite photos to prove NATO or Washington claims. A *Sputnik* report of February 20 cited Russia's Defense Ministry's treaty control department as asserting that Dutch military experts for OSCE had detected no Russian military activities in the Rostov region bordering Ukraine (Boiko 2015).

The Minsk II ceasefire agreement was signed by the second week of February. Ukrainian troops were being hammered by separatist forces generally and in particular at Debaltseve, which the separatists had taken in April and which Ukrainian forces were trying unsuccessfully to retrieve. Defeat was imminent. It was this that had sent the German premier Merkel and the French premier Hollande to see Russian President Putin, "the man they and others kicked out of the

G8 group and labeled a tyrant" (Kimberley 2015). Point 10 of the Minsk agreement required that "all foreign troops, heavy weapons and mercenaries were to be withdrawn from Ukraine. Illegal armed groups would be disarmed, but local authorities in Donetsk and Luhansk would be allowed to have legal militia units" (McAdams 2015). This seemed to point to regular Russian troops that Ukraine had consistently claimed, without compelling evidence, to be fighting in the Ukraine and whose presence Russia denied. But it would also apply to US troops, the CIA and Special Forces. Why would it not apply, therefore, to the 600 US trainers committed to Ukraine by Washington and who arrived early in March? Sources in mid-March were conflicted as to whether the trainers had actually been sent by the forecast date of arrival or whether, if they had arrived, they had started work in the field, but they were reported as being in close proximity to battle zones in April.

## Tensions within the Western coalition, contested narratives

At this very point, following Minsk II and amidst continued indications that the USA, the UK and some other Western governments would be sending personnel and "non-lethal" assistance to Ukraine, Germany (in particular) and the EU more generally began to publicly clash with what they considered was the unhelpful, continuing interference from the USA. Most notably, German officials made it known that US officials who regularly claimed there were ceasefire violations and a Russian military presence in Ukraine were engaging in "dangerous propaganda" meant to sabotage the ceasefire. The US claims, said these German sources, were often wildly inaccurate or unconfirmed as on the occasion when the USA claimed in April 2014 that "40,000" Russian troops were at the border and an imminent invasion was expected, when by contrast EU intelligence showed only 20,000 troops, and that no invasion was likely. And when in mid-February 2015 the USA had claimed that 50-plus Russian tanks, forty missile systems and forty armored vehicles crossed the border, the USA was simply lying, said German intelligence sources. The only evidence was a handful of ramshackle armored vehicles crossing the border, which likely were not even military in origin (Ditz 2015a). A few days prior, the German magazine *Der Spiegel* reported General Philip Breedlove, the top NATO commander in Europe, complaining that Putin had once again "upped the ante" in Eastern Ukraine – with "well over a thousand combat vehicles, Russian combat forces, some of their most sophisticated air defense, battalions of artillery" sent to the Donbass. "What is clear," Breedlove said, "is that right now, it is not getting better. It is getting worse every day" (*Der Spiegel* 2015). The German government, supported by intelligence gathered by the Bundesnachrichtendienst (BND), Germany's foreign intelligence agency, certainly did not share this assessment. There was a regular pattern, *Der Spiegel* noted, in which Breedlove would speak of troop advances and the amassing of munitions and columns of Russian tanks, while his numbers were found to be significantly higher than those in the possession of America's NATO allies in Europe. NATO has no intelligence agency of its own but depends on intelligence gathered by agents from the USA, Britain, Germany and other member states. When it wasn't Breedlove's dramatic evidence-free assertions of Russian aggression, it was the turn of General Frederick Ben Hodges, Commander of US troops in Europe who, early in March, accused Russia of having 12,000 troops inside Eastern Ukraine. But even if these fear mongering claims were correct, to whom were they a matter of concern? As one commentator noted:

> The US-NATO coalition willfully ignores the fact that Ukraine is not a member of either the European Union or NATO and has no treaty of any sort with any nation in the world that would require provision of political, economic or military support in the event of a bilateral dispute with any other country. Yet NATO has seized upon the Ukraine-Russia discord to justify its policy of unrelenting hostility to Moscow. NATO should have been disbanded at the time of the collapse of the Soviet Union because that threat was the sole reason for its existence; but it decided to multiply membership and extend its military presence closer and closer to Russia's borders.
>
> (Cloughley 2015)

Further evidence of Western and Ukrainian overstating Russia's presence in the Ukraine, if any, was suggested, paradoxically, by advance intelligence of a report to be released by supporters of the assassinated Russian opposition leader Boris Nemtsov. It had been mooted that one of the reasons for his assassination was his work preparing a report of how Russia allegedly discharged its soldiers so that they could fight in Ukraine as volunteers. When these were killed, Russia reportedly refused to pay compensation to their families in an attempt to cover-up official involvement. Two waves of such volunteering were alleged: one in the summer of 2014, and the other in fighting that took place in January and February of 2015, prior to the Minsk II agreement. In the summer, however, only "scores" of such volunteers were reported; while in the second wave, there were allegedly seventy deaths of volunteers. In both cases the numbers are relatively small, and certainly not of the order of magnitude that would be expected of an actual invasion. Their purpose in the summer of 2014 was said to be to secure an advance by separatist forces, a role akin to "advisers" that were likely then and later sent by the West in support of Ukraine (Dolgov 2015). In any case, the evidence that such volunteering was in fact a mandatory imposition on soldiers is still far from watertight. Lesser forms of Russian, though not necessarily State-sponsored Russian assistance to the militia in Eastern Ukraine included crowdfunding campaigns such as Save the Donbass, Veche, Novorossiya Humanitarian Battalion, making use of the Internet, YouTube and social media such as VKontakte (Becker and Myers 2015).

Shortly after the Nemtsov revelations in April 2015, senior French intelligence sources debunked NATO warnings of a Russian invasion of Ukraine, and claimed that these were "based on stove-piped intelligence fed to NATO largely by Washington." Western media, including French, ignored public testimony at the French National Assembly by Director of Military Intelligence General Christophe Gomart. Gomart complained that NATO intelligence was preponderantly fed by the USA (Hugues 2015).

French intelligence had found nothing to support NATO claims – Russia "had not deployed command centers, a supply chain or military hospitals, and reserve units had not moved at all, even though individual Russian soldiers may have been spotted in Ukraine as a form of Russian pressure, Gomart alleged, on the Ukrainian president. Gomart's assessment was supported by Philippe Migault, an East European expert at the French Institute of International and Strategic Relations think-tank (Hugues 2015). Writing for the *World Socialist Web Site*, Stéphane Hugues commented:

> Yet it appears that false information largely provided to NATO by Washington was used to press for NATO preparation for an imminent Russian invasion of Ukraine – that is, a major military build-up placing NATO military forces across Europe on high alert aimed at Russia. The risks posed to the world's population by such a criminally reckless policy, which heightens the danger of the eruption of all-out war between NATO and nuclear-armed Russia, are incalculable.
>
> (Hugues 2015)

Increasing German and French concerns about NATO misinformation may have contributed to a distinctly lukewarm attitude to Ukraine and the supposed Russian threat during the European Union summit in Riga on the Eastern Partnership, May 21–22, 2015, where the EU committed itself only to a provisional application of the Deep and Comprehensive Free Trade Agreement (DCFTA) with Ukraine which was due to start on January 1, 2016. The following G7 meeting in early June made routine threats of continuing sanctions on Russia if Russia did not put a stop to violations of Minsk II (as though this was something that Russia could control) and perhaps even of increasing sanctions in future, but did not mention Crimea. In September, the EU extended by six months a visa ban and asset freeze targeting several close associates of the Russian President Vladimir Putin and others involved in Russia's annexation of Crimea and other territorial disputes in Ukraine.

Furthermore, at the end of June 2015, Gazprom (which has been variously reported to account for 25 percent of Russian income or 8 percent of Russia's GDP) announced plans to extend Nord Stream pipelines in cooperation with several European oil companies, enhancing Gazprom's and Europe's ability to sideline Ukraine as a country of transit. Nord Stream had commenced in 2011 and by 2015 the European oil companies working in collaboration with Gazprom included OMV (Austrian), E.On (German), British-Dutch Royal Dutch Shell, Wintershall (German) and GDF Suez. Gazprom also collaborated with Shell in Sakhalin-2 (LNG) in the Pacific. The extension of Nord Stream provided European companies with an opportunity both to exploit Russia's negotiating weakness as a partner, and to enhance European strength against the USA (Weiss 2015a). On the other hand, a massive build-up of US LNG was anticipated in 2017 and 2018, together with new supply from Australia and Qatar.

The discourse of the Russian "invasion" of Ukraine sidelined a more fundamental issue of Russia's long-standing penetration of the Ukrainian economy, operating in part through its influence in the security service, SBU and in the economy which it subsidized to the tune of at least $5 billion a year. But more relevantly in this context it sidelined increased evidence of the presence in Eastern Ukraine of 600 US, UK and Canadian military troops – "advisers" – working with the National Guard in and around the towns of Artemivsk, Severodonetsk and Volnovakha only 20–30 kilometers from the conflict zone. All this was in the context of Ukraine still retaining heavy artillery and engaging in practice of live-fire drills close to the conflict zone. Several skirmishes and deaths had been reported. In mid-April, RT quoted OSCE reports of a spike in violence near Berdyansk and Shirokino. During this period, the Russian Foreign Ministry complained about the Western build-up of forces in Estonia, Latvia and Lithuania, something that it said was forbidden by the Russia-NATO Founding Act of 1997, still in force. NATO was also in discussion with Romania to house a missile defense system in the town of Deveselu (RT 2015d). In the meantime, the US ambassador to Ukraine in Kiev, Geoffrey Pyatt, and the US State Department, stepped up "Russian invasion" propaganda claiming a Russian military presence in Eastern Ukraine (supported by a photo of a Buk missile, lifted from a Russian exhibition two years previously) and Russian installation (no evidence supplied) of advanced air defense systems (Cunningham 2015c).

Ukrainian claims of a direct Russian presence heated up in May 2015 at the same time as post-Minsk II fighting returned to significant levels. Reuters reported a separatist attack on the Ukrainian (and Azov) controlled town of Marinka, west of Donetsk on June 3, which led to the wounding of twenty-five Ukrainian soldiers (Reuters 2015d). Ukraine claimed to have captured two Russian servicemen of whom one, on video, said he had been part of a fourteen-member group (Reuters 2015c). They were interviewed by UN officials who concluded that it was very difficult to prove whether they were servicemen or not. Early in June, senior UN human rights officials declared increasing evidence, but not yet conclusive legal proof, of Russian state involvement, in the Donbass (UNIAN 2015a). Such incidents occurred on both sides: in September in an incident not picked up in the Western press, border guards of the Federal Security Service of the Russian Federation reported they had detained three soldiers of the 79th brigade of the Armed Forces of Ukraine in Crimea suspected of illegally crossing Russia's state border (UNIAN 2015h).

According to a report emanating from US propaganda source RFL/RL and published in the *Guardian*, a volunteer army of Ukrainian bloggers was claiming to have found evidence of social media posts and comments on Russian-language websites from Russian soldiers posted to Eastern Ukraine (Volchek and Bigg 2015). Referencing a number of US press evidence-free, speculative scare stories in late May, William Boardman (2015a) identified deliberate confusion on the part of NATO authorities as to whether Russians had actually invaded or were amassing on the border (no evidence supplied for either), a very important distinction. According to one anonymous source:

> Russia had moved about nine battalion tactical groups close to its border with Ukraine, and that as many as five additional battalions could be sent there in coming weeks. The number of troops in such units can vary, but a battalion could have about 1,000 troops, creating a potential force of well over 10,000 Russian troops in Ukraine by this summer.
>
> (Boardman 2015a)

The report to which Boardman referred was by Michael Gordon (culprit, with Judith Miller, for helping scam the USA into invasion and occupation of Iraq in 2003) of the *New York Times*. Other than outright propaganda, available evidence suggested not an overwhelming presence of Russia in the Donbass, but some mixture of support, including, possibly, military personnel. The moving report of one visitor, Joshua Tartakovsky, who traveled to Donetsk in mid-May confirmed overwhelming civilian support for the independent republic of Donetsk and corresponding distrust of what was perceived as the Bandera-supporting fascist government of Kiev (Tartakovsky 2015). There was disturbing evidence of constant shelling of the civilian population by the Ukrainian military. Evidence of such shelling was also supplied from interviews with earlier refugees from the Donbass in camps on the Russian side of the border (Mokrushyna 2015a). On the presence of Russians, Tartakovsky was ambivalent: his report appears to: (1) deny a substantial Russian presence; (2) confirm the broadly local character of the fighters; and (3) concede a likely presence of Russian soldiers, without determining whether they were regular or volunteers.

> While Russia undoubtedly lends its support to the self-proclaimed republic via various means, it would be incorrect to say that the local population does not stand behind the republic. [...] Throughout the city, we did not encounter masses of Russian soldiers marching in perfect order, nor columns of Russian tanks. The fighters we encountered were locals, while we were told that a sizeable number of international volunteers came from Spain and Germany. A significant number of soldiers must have come from Russia. We later saw various tanks and Armed Personnel Carriers with flags of the DPR (Donetsk People's Republic). We were forbidden to take pictures of military.
>
> (Tartakovsky 2015)

Tartakovsky's impressions were very different to those of pro-NATO Atlantic Council's senior fellow Adrian Karatnycky whose evidence-free picture of the Donbass was one of criminal pro-Russian warlords undertaking Putin's dirty work and even in danger of taking over from him: leaders of gangs, jobless males, pro-Russian extremists of the far right (including Cossacks) and mercenary veterans of the Chechen and other Russian wars (Karatnycky 2015). Their revenues, Karatnycky claimed, came from trade in weapons, drugs and alcohol; checkpoints; shakedowns of local businesses; and seizure of abandoned homes. Karatnycky's perspective was supported by a story in the pro-Kiev Ukrainian news agency, UNIAN, citing the Kharkiv Human Rights Protection Group and the Russian newspaper *Novaya Gazeta.* This claimed that four individuals, with top political and/or military connections, effectively controlled all business in the Donetsk region: Aleksandr Zakharchenko (leader of the DPR); Aleksander Khodakovsky (head of the DPR's Security Council, and enjoying the support of ex-Yanukovych sponsor oligarch Rinat Akhmetov); Denis Pushilin (deputy prime minister) and Igor Girkin (implicated as a potential culprit in the case of MH17), former deputy prime minister and head of the Novorossiya army who may also have worked for Russian military intelligence GRU (UNIAN 2015o).

A report by Valentyn Nalyvaichenko, ex-chief of the Security Service of Ukraine (SBU) until his falling out with President Poroshenko, and leaked to Bloomberg in early July, featured various examples of Russian activities aimed at undermining Ukraine's sovereignty, from espionage to supplying military vehicles and Russian troops, and claimed there were 8,960 Russian regular troops and 33,400 Kremlin-backed separatist troops in Donbass. The report was allegedly prepared for the US government (Mikhailov 2015). Around this time senior UN human rights officials declared increasing evidence, but not yet conclusive legal proof, of Russian state involvement in the Donbass. Weak supporting evidence was supplied by none other than Russian President Vladimir Putin himself during his annual end-of-year news conference in Moscow, on December 17, 2015. He was asked by a Ukrainian reporter about two captured Russian military intelligence officers on trial and Putin responded: "We never said there were not people there who carried out certain tasks, including in the military sphere," but insisted that these were not regular Russian troops (*Moscow Times* 2015b). In December, Ukraine intelligence reported seven Russians killed in action and twelve wounded in recent fights in Donbass.

The report of the UN Human Rights Monitoring Mission in Ukraine in September 2015 complained that the situation "continued to be marred by ongoing armed hostilities in some areas of Donetsk and Luhansk regions reportedly fueled by the presence and continuing influx of foreign fighters and sophisticated weapons and ammunition from the Russian Federation" (Williams 2015b). Use of the term "reportedly" indicated continuing confusion at the uppermost levels of the international organization as to the reality on the ground. Further the wording appears biased against the interests of ethnic Russians in the Donbass and Russia, failing to distinguish between aggressive and defensive measures or to entertain the consequences of allowing free rein to a Ukrainian army known to comprise fascist elements. This bias was evident in the report's strictures on separatists' "development of more centralized civilian administrative structures and procedures," a view that failed to consider the withdrawal by Ukraine of pensions, social security and other central services from the Donbass.

While anti-Russian sources made great play of whatever evidence they could find of the presence of Russian volunteers fighting in defense of separatists in the Donbass, Western media studiously ignored evidence of 1,000 or so anti-Putin volunteers from Russia, Belarus and elsewhere fighting on behalf of paramilitary forces against the separatists. Such foreigners included those recruited in 2014 by Semen Semenchenko, commander of the Donbass Battalion and later a member of Ukraine's Parliament. The Battalion was a volunteer paramilitary unit fighting against Russia-backed separatists in Eastern Ukraine before it was later incorporated into the Ukrainian National Guard, assisting the Ukrainian Army in Luhansk and Donetsk. The Interior Minister Arsen Avakov hailed the foreigners as heroes and promised them "fast-track citizenship" a promise that was not quickly fulfilled, leaving many stuck in Ukraine without legal status. In addition, neo-Nazis were recruited from as far away as Sweden, and had been openly active in Azov, an ultra-right outlet. One informant estimated there were 100 such foreigners in Azov (Antonova 2015).

## Post-Minsk II ceasefire

In July 2015, top officials from the Donetsk and Luhansk People's Republics officially announced that they had taken the unilateral decision to withdraw at least 3 km from the front line all their weapons up to a 100 mm caliber (weapons of a heavier caliber were supposed to have already been withdrawn according to the Minsk-II Agreement [M2A]). Previous unilateral action had withdrawn all forces by over 1 km from the town of Shirokino. Ukraine initially responded to the initiative with another barrage of missiles shelling Donetsk. The Republics had found it increasingly difficult for the ethnic Russian forces to show the West that it was the Ukraine, not they, that were in continuing violation of the Minsk II agreement. One analyst predicted a Ukrainian attack that would be repulsed by the ethnic Russian forces, leading to the toppling of Poroshenko and, under pressure from Right Sektor, the carving up of Ukraine into many smaller units (The Saker 2015). However, on July 21 the OSCE Chairman-in-Office, Serbia's First Vice Premier, Minister of Foreign Affairs Ivica Dacic announced that smaller-caliber weapons and tanks would be withdrawn from the contact line in Donbass in Ukraine's east. The preliminary deal reached by the two sides focused on weapons smaller than 100 mm (the larger ones having already been withdrawn, supposedly).

Certainly the situation remained unstable over the summer of 2015, although accounts of skirmishes were generally over-exaggerated. In August, Kiev claimed that hundreds of pro-Russian fighters, tanks and heavy artillery had attacked Ukrainians near the small town of Starohnativka. Ukrainian intelligence, SBU, released the names of several high-ranking Russian officers who were leading the "militants" in the Donbass. The attacks were denied by separatist authorities in Donetsk who accused Ukraine of making up the story and insisted the separatist army had not broken the ceasefire. According to Jack Losh, reporting for the *Independent*, eyewitness accounts from the front line on either side of the divide challenged both official versions of events, dismissing the claims and counter-claims as propaganda to lift morale (Losh 2015a). OSCE representatives noted that both the Ukrainian side and the self-proclaimed republics had failed to observe the agreement on the withdrawal of heavy armaments.

Reports of ceasefire violations persisted through to the end of 2015, amidst otherwise encouraging signs of conformity with Minsk II, and withdrawal of both small and heavy arms. This considerably reduced the pressure on Ukraine's army. Early in October, the Russian agency TASS cited Ukraine's Chief Military Prosecutor as saying that a total of 16,000 Ukraine soldiers had defected during the war, many of them with weapons. During a sixth wave of call-up almost 27,000 conscripts (almost 50 percent of the total) had dodged service (TASS 2015). There was an escalation of claims from both sides during November 2015, with reports of deaths and shelling in the Luhansk and Donetsk regions, with some indications that these would serve as a pretext for returning arms to previously cleared areas or justifying the surreptitious retention of such arms. For example, towards the end of November, Ukrainians complained of separatist fire from the city of Donetsk involving, it was claimed, 120 mm mortars, anti-aircraft guns and grenade launchers for firing in the direction of the town of Krasnohorivka and villages of Pisky, Opytne and Novhorodske, and the town of Avdiyivka (UNIAN 2015o). The Organization for Security and Co-operation in Europe (OSCE) confirmed that the Ukrainian military had moved heavy weapons back to the contact line.

## The chances for decentralization, the October 2015 elections and Ukraine media

In the West, some journalists noted a cooling-off in attitudes of European governments and European media towards the USA, in favor of a better understanding with Russia (Tikhonova 2015). In a possible sign of disenchantment by the EU with US and NATO warmongering, the French, German and Russian governments brokered an agreement between Ukrainian and ethnic Russian separatist forces on August 26 that from September 1 they said would strive for an end to all true violations, and that there would be a summit within weeks. This took place when the Ukrainian, Russian, French and German Presidents met in Paris early in October. These endorsed the elections scheduled for October 25, but questioned ethnic Russian organization – in the face of Kiev disapproval – of elections in the Donetsk and Luhansk People's Republics scheduled for October 18 and November 1 respectively. There were reports (Xinhua 2015, citing the Ostrov News Agency) in Donetsk that the leaders of the Donetsk People's Republic planned a referendum within two weeks of the October 18 election asking citizens to vote on joining with Russia (Xinhua 2015). The republics later postponed the dates for these elections, pushing them forward to 2016. The Paris talks indicated that Russia was pushing Ukraine to take back legal control of the Donbass, to grant amnesty to the separatists and legalize elections run by separatist forces. Ukraine was unwilling to make such concessions without securing full political control (Walker 2015).

The separatist administrations – in a clear indication of a federalist as opposed to a secessionist agenda – said elections would be held only after Kiev assigned their regions "special status" within a unified Ukraine with the right to develop closer diplomatic and trade ties with Russia. Under Minsk II, Ukraine was obliged to grant more autonomy to regional governments to allow them to decide their own economic and foreign policies. A bill to this effect was debated and passed by 265–185 on a first vote by the Supreme Council in August 2015 (only a bare majority was needed), but faced strong opposition from nationalist deputies and amidst violent demonstrations in Kiev, orchestrated by the Svoboda and Radical parties, and Right Sektor militia, that injured over 140 people and killed four. Most of the victims were policemen. Even this bill was considered by some experts to be more centralist than decentralist, with many legitimate problems of wording, interpretation and concerns, in particular, about the role and powers of prefects that would be appointed by the President (cf. Mokrushyna 2015b). At the time of writing, the bill still awaited a second reading, which would require a two-thirds majority. Separatists also sought full immunity from prosecution, together with a new vote in the Ukrainian parliament on constitutional amendments regarding elections that would first be agreed with the rebels themselves. Under the Minsk II agreement, Russian forces and militias would cede a 250-mile stretch of the border of Russia with Ukraine back to Ukraine. But separatists would only do this once Kiev followed through on all the commitments it had made in Minsk II (Silino and Shylenko 2015).

The first round of voting on October 25, one observer concluded, was marked by widespread accusations of fraud and intimidation and relatively low voter turnout (46.6 percent, falling to 36 percent in the second round in November). The 1.5 million internally displaced people were not allowed to vote, many of them on the Ukrainian edge of the 15-mile buffer zone between the Donbass and the rest of the Ukraine, created by the Minsk II treaty, and where the Opposition Bloc was likely to have done quite well. Here, thousands of houses lay in ruins, many settlements lacked utilities, pensions stood at $165 a month and many householders had to put up with Ukrainian soldiers forcibly sleeping and feeding in their houses (Ferris-Rotman 2015). Observers from OSCE nonetheless approved the conduct of the voting. Polls conducted by the Razumkov Centre shortly before the election showed that almost half of Ukrainians believed that no party represented their interests (UNIAN 2015i). According to a national survey conducted in August 2015 by the Ukrainian Sociological Group "Rating," commissioned by the International Republican Institute and funded by USAID, an absolute majority of Ukrainians were dissatisfied with the Verkhovna Rada: 49 percent of Ukrainians did not approve at all of its activities – one percent short of the dissatisfaction with the Verkhovna Rada of February 2014 – when it was considered to be anti-Maidan, while 35 percent are "rather dissatisfied." Fifty-two percent of Ukrainians did not approve of Prime Minister Yatsenyuk and his Cabinet of Ministers at all, while 32 percent tended towards not approving. Thirty-three percent did not approve the actions of President Poroshenko, while 34 percent "rather do not approve."

The situation had worsened by the end of the year when a Gallup poll showed Poroshenko rating less popular than his predecessor Viktor Yanukovych before he was ousted: only 17 percent of Ukrainians approved of the job that Poroshenko was doing. This approval rating was down sharply from 47 percent a few months after Poroshenko's election in May 2014. Poroshenko was not popular in any region of Ukraine, least of all in the East and South. Only 8 percent expressed confidence in their government, only one-third of where this measurement had stood in 2014. The large majority considered corruption was rampant in government and business (Ray 2015). Overall, concluded Mercouris,

> Ukrainians are deeply dissatisfied with the state of current affairs and with the country's governance. [...] Ukrainians are dissatisfied with the government, with the president, with the economy, with the reforms, with their own economic situation. The euphoria of Euromaidan is over for many.
>
> (Mercouris 2015b)

Even though the Communist Party, once Ukraine's biggest party, had been banned and the Opposition Block had been banned in parts of Western Ukraine, with attempts made to ban it in Kharkov, the October election results demonstrated the weakness of the parties of both the president and prime minister. Voting was canceled at the last minute in the eastern cities of Mariupol, Krasnoarmeysk and Svatove, possibly because of fears of a win by the Opposition Bloc. Elections did not take place in two towns in Eastern Ukraine – Krasnoarmeysk and the key port city of Mariupol, perhaps to prevent sweeping victories by the Opposition Bloc. In Kiev, the biggest opposition newspaper had been closed after months of harassment and a series of murders of opposition politicians and journalists. Elections were not held in the cities of Avdiyivka, Maryinka, Artemivsk and Kostiantynivka. The Opposition Bloc won sweeping victories in those parts of Southern and Eastern Ukraine where the elections took place. The Opposition Bloc did surprisingly well in Kramatorsk, the de facto capital of what was left of the Donetsk region. In Kharkov, Gennady Kernes, a former Party of the Regions politician, retained the mayoralty by a landslide. The party headed by Ukraine's Prime Minister Arseniy Yatsenyuk had declined to take part in the elections at all, likely fearing a wipe-out. President Poroshenko's party nationally fell to around 18 percent. His support was increasingly concentrated in the small towns and regions of central Ukraine. In Odessa, Saakashvili – the man Poroshenko had appointed to run it – was conclusively defeated. The neo-fascist party Svoboda did well in Lviv and Kiev (Mercouris 2015b).

The performance of Ukrainian media was a factor that deserved greater attention by WMM, given their typical skepticism as to the quality of Russian media coverage. In the post-election assessment of OSCE's Office of Democratic

Institutions and Human Rights, for example, OSCE noted that "political and business interests controlling the media often influence editorial policy, and paid-for coverage is widespread" (Rudenko 2015). Another source chronicled widespread corruption in Ukraine's television channels and other media outlets, beginning with President Poroshenko's ownership of Channel 5 known for its "laudatory coverage of the government" and

> positive stories that personally feature Poroshenko and frequently produces uncritical items about the president's routine activities, such as statements issued by his office or details of his travels, leading media monitors to conclude that it is "slowly turning into Poroshenko's press office."
>
> (Fedets 2015)

All ten of the country's most popular channels were owned by powerful oligarchs: three by Victor Pinchuk, three by Ihor Kolomoiski, three by Dmytro Firtash and one by Rinat Akhmetov. All "use their media might to advance their business and political interest [...] most of the country's top TV channels air political advertising promoted as 'news'" (Fedets 2015). The oligarchs managed media outlets through numerous companies so that viewers were often unaware who controlled the editorial line of a given channel.

Nationwide voter turnout was at a low of 46.6 percent. The lowest participation was in those disputed eastern districts of Donetsk and Luhansk that were still under Kiev government control, where 32 percent and 35 percent of eligible voters, respectively, cast ballots (Williamson 2015b). There were some confrontations over the mayoral elections of October 25 in Mariupol (where the election was cancelled because of fear of fraud) and Kharkov. In Kharkov, 200 members of an Azov battalion were reported by the local television channel Ukrainian 112 to have occupied a local administrative building, in a bid to oppose the Opposition Bloc political party (RT 2015g). About 5,000 demonstrators protested on November 22 against alleged voting fraud in the November 15 mayoral run-off election in Dnipropetrovsk Oblast's Kryvyi Rih. Official results favored the incumbent Mayor Yury Vilkul, a former associate of the ex-President Viktor Yanukovych. A Dnipropetrovsk court rejected all complaints. Protestors occupied the City Hall and sought to involve Kiev authorities in a dispute over the result (Sukhov 2015b). Ultimately, new elections were called for March 2016. In Mariupol, a city of 500,000 residents, the election was rearranged for November 29, amidst fears of attacks by separatists and Right Sektor patrols on polling stations. Turnout for the first stage of the vote was 35 percent (Grytsenko 2015).

In this chapter, I have argued that contrary to the presumptions of WMM and Ukrainian media and propaganda as to the aggressive intent of Russia and the presence of Russian troops and military support for the ethnic Russian separatists of the Donbass, the situation on the ground was persistently difficult to establish with confidence. There is significant evidence that NATO and other Western sources deliberately sought to escalate tensions by gross exaggeration of Russian intervention in Ukraine. Furthermore, any Russian presence likely fluctuated in form and intensity over time but never at any point approached the degree of intervention proclaimed by NATO military leaders.

## Note

1 This chapter builds on Oliver Boyd-Barrett (2015): "Ukraine, Mainstream Media and Conflict Propaganda," *Journalism Studies*. Reproduced with permission of Taylor and Francis.

# 10 Articulating the principles of Western conflict propaganda

Propaganda functions variously to change or reinforce beliefs, attitudes towards and values about agencies, people, events and things (Jowett and O'Donnell 2015). Typically in Western media, propaganda does not announce itself as such but, as Edward Herman and Noam Chomsky and many others have long argued, it achieves the purposes of the propagandists through the ways in which issues are framed, through emphasis and omission, the privileging of certain sources, perspectives, information over possible alternatives, and in the uses of language (verbal and visual) that help achieve these effects (Boyd-Barrett 2015; Herman and Chomsky 2002). In this paper, I am choosing to focus on what appear to me to be some of the understandings (propaganda claims) about Russia, Ukraine and the West that the propagandists – the Western powers whose voices, values and preferences are privileged by Western media – hope will prevail among members of the public. My focus in this book has been primarily Western and WMM propaganda. Clearly, other parties including Russia, Ukraine and the Donbass also engage in propaganda – but they are not my main focus.

I identify different preferred understandings – the desired propaganda belief – by their respective principal topics.

## Allies

If an ally whom the West has shooed into power by unconstitutional means considers itself threatened by a powerful neighbor, then the West should protect its ally regardless of whether or not there are any treaties that require or legitimate this. If on the other hand the more powerful neighbor considers itself threatened because the West has circled it with missiles, has – despite previous undertakings to respect the neighbor's legitimate security interests in the region – penetrated militarily, in this case through NATO, closer and closer to the neighbor's borders, supports its smaller ally's aggressive attacks on those of its own people who consider themselves ethnically close to the neighbor, offers lethal and non-lethal assistance to its ally against the neighbor, then any indication (potential or actual – it hardly makes a difference) that the neighbor will seek to secure its borders, protect its ethnic brothers and sisters, and ready itself in preparation for potential further Western aggression is considered heinous warmongering and imperial expansionism that the West cannot tolerate.

## Attack by an army against its own people

If a regime that is favored by the West deploys its army against its own people, as happened when the authorities of Kiev attacked and ravaged the separatist movements of the Donbass, then that is perfectly excusable and is also a pretext for Western aid to that regime against regime opponents. If the regime is not favored by the West, as in the case of Saddam Hussein's Iraq, Muammar Gadhafi's Libya or Bashar al-Assad's Syria then evidence of such deployment, whether real or fabricated is an acceptable pretext for Western humanitarian intervention and democracy promotion whose purpose is regime-change.

## Attribution of good and evil

Required to choose between a benign or a malign interpretation of Russian activities as these might relate to the conflict between separatists, Ukrainian army, Ukrainian militia and Russia, the West generally privileges the malign interpretation.

## Context

Western media's sourcing of stories of alleged Russian aggression and imperialistic hubris in the Ukraine require no contextual referencing to recent Western instigated invasions and occupations of Afghanistan and Iraq, disastrous Western instigation of and/or participation in the destabilization or attempted destabilization of Libya, Syria, Yemen and Iran, and a history of US-backed "color revolutions" in the former Soviet Union and elsewhere designed to promote US interests under the banner of democracy building. Nor does contextualization of the unrest or civil war in Ukraine require reference to very long-standing ties (1,200 years) between Russia and Kiev, the substantial presence of Russian language and Russian-speakers in Ukraine, a history of at least one hundred years of Western aggression towards Russia, the West's violation of its own agreement in 1990 with ex-Russian President Gorbachev that NATO would not co-opt the former East European states and thus expand its influence to Russian borders, evidence of US strategies to impede alliances between Russia and Western European powers or between Russia and BRICS powers, US competition with Russia for the resources of a thawing Arctic, US competition with Russia for control over Eurasia, the relevance of US/NATO strategies for the destabilization of Syria and Iraq to their longer-term significance for the destabilization and fragmentation of the Russian Federation, and to the USA's struggle for hegemony of its prevailing petrochemical order.

In covering the official claims of Western governments as to their policies towards countries that have been targeted for co-option into the NATO security sphere, such as post-coup Ukraine, including condemnation of Russian "aggression" and support for democracy, Western media need not consider as relevant the long-standing weaknesses of these countries (including in the case of Ukraine its long-standing economic basket-case status) and the contribution to them of the excesses and non-democratic instincts of their oligarch classes. Nor need they consider as relevant to Western motivations the loot for Western corporations that are consequent on the IMF and other Western loans, involving as these typically do the privatization of infrastructure, shipping and rail, and the legalization of market access for foreign interests in traditional and GMO agribusiness and other spheres. While Western media may feel entitled to present Russia's Gazprom as a villain, sucking up Ukrainian wealth by the imposition of unfair prices, they need not consider Gazprom as a source of subsidy to Ukrainian energy consumption, given its reluctant dependence on Ukraine as a transit for the supply of its gas to Europe. Nor need they attend to the machinations of Western energy companies vying for advantage on Ukrainian markets or of the US and EU attempts to derail Russian efforts at establishing alternative oil and gas supply routes, and at destabilizing the Russian economy by such means as alliances with Saudi Arabia for price-cutting.

## Democracy

A state that is largely run by oligarchs and whose democracy is based on political parties that are largely affiliated with the interests of different oligarchs (e.g., Ukraine) should be supported by Western powers so long as it can exhibit the trappings of modern democracy such as regular elections, diversity of choice etc. and even if the democratic process must be completely upturned in order to bring to power a government that is acceptable to Western interests. A state that has many oligarchs but whose power nonetheless is contested by a strong executive within a democratic, multi-party system cannot be regarded as an acceptably functioning democracy when Western powers don't like the interests that it represents or advances.

## Elections

Elections for independence that take place under the rule of authorities who do not enjoy the approval of Western powers will be considered less than legitimate by those powers and their media, especially when they occur in periods of conflict and unrest (as in Crimea). Elections for independence that take place under the rule of authorities that are approved by Western interests may not even be necessary (as in Kosovo) or, if held, are reassuring signs of democracy even in the most unpromising of circumstances (as in the presidential elections in Ukraine that elected Poroshenko to the presidency, and the parliamentary elections that were held in August 2014).

## Evaluation of historical events

Historical events that are deemed to have had negative implications for the possibility of autonomy of a nation which currently enjoys Western support are bad, and whoever had responsibility for such events acted with malign intent (e.g., Western Ukrainian evaluation of Soviet collectivization in the 1930s). Historical events that are deemed to have had positive implications for the possibility of autonomy of a nation which currently enjoys Western support are largely good or can at least be tolerated, as in the case of Stepan Bandera's fascist Ukrainian Insurgent Party in the 1940s.

## Evidence

If an "eyewitness" claim is made by a journalist from the mainstream Western media, possibly supported by a companion journalist from an equivalent mainstream media news publication, and it is not backed up by any evidence other than for the claim itself, then the claim can be assumed to be true and other Western media are entitled to pick up and re-assert that claim without further investigation, even though the meaning of what was claimed to have been seen is less than clear. This was the case with Western journalist claims to have witnessed the incursion of a Russian military convoy into Ukraine in August 2014.

If a spokesperson for the authorities of a party to whom Western powers are favorably disposed asserts a claim, without further evidence, then that claim may be assumed to be true, and the claim may be picked up and re-asserted by those at the very highest levels of the party or of the Western powers which support it. This was the case when Ukraine claimed to have destroyed part of a Russian convoy in Eastern Ukraine in August 2014.

Where official evidence in support of Western claims of Russian aggression is lacking, it is perfectly acceptable for Western media to employ the assistance of self-appointed "citizen journalist" technology sleuths with possibly if shadowy connections to the Western intelligence establishment and to place their claims on a par with official sources (as in the case of the *Guardian*'s use of Bellingcat.com to determine the trajectory of missiles).

Social media, highly prone to corruption, are also perfectly acceptable sources of intelligence, never less than when touted by US Administration figures who must assuredly have much stronger and more direct intelligence evidence available to them. Social media trumping of official evidence was illustrated by John Kerry's statements on MH17 in the immediate aftermath of the tragedy, fingering Russia as the culpable agent, which cited social media rather than the direct evidence of what must have been hundreds of intelligence agency reports (McGovern 2015).

Evidence from the media of opposing non-Western powers (e.g., RT) cannot be trusted because such media act as mere channels of propaganda, except when media of those powers happen to report something that confirms the Western narrative of events – as in the case of the *New York Times* reporting a story from

Moscow's *Novaya Gazeta* that in turn originated from a dubious source for which there was no supporting evidence.

Although the evidence of recent Western expansionism in all of its various forms is far more voluminous and compelling than evidence of comparable Russian expansionism, Western accusations against Russia should be taken at face value and do not require contextualization with reference to recent Western expansionism.

When the US government provides or claims to have in its possession photos or satellite images as evidence in support of its narratives, it can be taken for granted that the photos actually do represent the events that the USA claims they do, or that the satellite images, if they indeed do exist, may be fairly interpreted the way in which the USA claims (despite egregious instances to the contrary). Such photo or satellite images, no matter their actual source, may be published without further investigation if made available by officialdom.

Finally, we can note how events that are initially reported by Western media as of uncertain status can be gradually converted to certain "facts" as a result, not of new evidence of any kind, but by dint of mere repetition and the slow simplifying conversion from ambivalence to unambiguity. There are many examples but perhaps the most egregious was the claim that Russia had invaded or was about to invade – the distinction seemed less important over time – Ukraine and that the separatists or independents in Eastern Ukraine were simply proxies for Russia, claims that conveniently suited intense Ukrainian propaganda but for which actual evidence was decidedly slender. The absence of evidence is in any case never an insurmountable problem for Western propaganda because it can always be construed as an indication of the devilish cunning of the enemy's capacity for subterfuge. This is another function of the demonization of the enemy: to make more devilish, more supernatural the enemy's power, including the power to make evidence disappear that Western propagandists "know" must really be there.

## Fascism

Political forces that subscribe to fascist doctrines such as racial purity and extreme nationalism, that craft a historical narrative that links them back in a line of succession to forces that assisted Nazi Germany's liquidation of Jews and Communists, that adopt iconic insignia that closely resemble those of Nazism, that conduct themselves against their enemies in barbarous, violent manner and directly threaten the democratic establishment when their wishes are not fulfilled, are not visible to Western propagandists and mainstream media when they are assisting in a *coup d'état* of which the West approves. Rather, voices that draw attention to such fascist elements are accused of greatly exaggerating their size and significance, aligning with Russian propaganda and overstating the threat. Victims of observed fascist violence, as in Odessa, are themselves responsible for the violence they have brought upon themselves, and their suffering merits no particular concern in Western media. On the other hand, ethnic

Russians fighting against Kiev forces in the Donbass are portrayed as in the clutches of and the inheritors of a tradition of Soviet communism, even though Soviet communism collapsed over a quarter of a century previously.

Dependence by a Western-supported coup regime on fascist forces backed by corrupt oligarchs, forces whose activities are not subject to control by the political and military mainstream, that are often criminal, and that have links with and support from radical Islamist movements, is apparently not a source of significant concern for Western governments or their media. Previously alleged criminal links of particular political and military leaders of the resistance to a Western-support coup regime however, as in the Donbass, are widely trumpeted as "evidence" of insincerity, corruption and inefficacy, and the entire separatist movement is smeared with frequently unfounded claims of its dependence on Russian support and assistance.

## Financial assistance and trade agreements

A political entity such as the democratically elected pre-coup regime in Kiev that does not proceed with an agreement for financial assistance from a source that is controlled or approved by the Western powers, such as the EU or the IMF but instead prefers the assistance of a source that does not enjoy such approval, such as Russia, is clearly acting on the margins of sanity, has probably caved in to unbearable and illegitimate pressure from the alternative source. It is unthinkable and therefore unspeakable – in the view of WMM – that backing-off from such an agreement with a Western-approved source of assistance could be the result of that source's intransigence in negotiation, or of its undue secrecy, or because the agreement actually was not in the best interests of the political entity to which the assistance has been offered or was simply not as good a deal as the one actually preferred.

## Humanitarian support

Actions taken to provide humanitarian support for the victims of an aggression committed by a party that enjoys the support of Western powers is inherently suspicious, treated as some form of military action in humanitarian disguise. This was the case with the Western reaction to Russia's preparations of aid convoys for desperate peoples in the Donbass.

## Invasions

When Western powers and their media claim that a country that they do not favor has invaded another country such claims do not need unambiguous evidence for support. Especially they need no evidence whatsoever of the scale of military action that commonly if not universally attends an "invasion" – for reference, see the US invasions of Afghanistan in 2001 and Iraq in 2003 – evidence that would be incredibly easy to provide if indeed such a scale of military activity had indeed occurred. "Invasions" may also be said to have occurred or obviously-about-to-occur if it can be claimed (unambiguous evidence not required) that the country in question has troops and equipment in the vicinity (but still on their side) of the border or even if there is some possible reason to assert that the country might at some point in the future act aggressively. This was what the West claimed Russia might do towards any and all of the pro-Western neighbors in the Baltics and Eastern Europe. These were firmly prodded by US/NATO/EU institutions into panic reaction and acting suitably afraid. Their compliance with these theatrical directions secured promises of and actual (further) Western military aid.

An invasion may also be said to have occurred even if the country's troops were already and legitimately present in a region of the "invaded" country (e.g., Crimea) a substantial majority of whose voting population expressed its fear of aggression from its "own" government – the coup regime in Kiev – and their desire to be taken under Russian protection. Once such an "invasion" has occurred (regardless of how weak the claim or how uninvestigated the actual circumstances) the cry of "invasion," pending or otherwise, can then be levied against the country in question whenever that country does anything, or perhaps nothing at all, that the Western powers do not like, so that the country in question continues to be vilified and its leaders demonized.

## Propaganda

For Western media, whatever statements are issued from official Western sources that lie close to the major sources of power, such as the White House, Pentagon or State Department are to be taken at face value, no matter how manifestly self-serving or reeking with hypocrisy, and should be accorded the priority that those sources would indicate their statements are worth. Such statements do not need to be independently verified in order to continue repeating them. Similarly, if at a slightly lower elevation, the statements, however manifestly self-serving or hypocritical, that are made by the allies of Western governments, such as Ukraine – no matter that country's recency, instability, problematic legitimacy, and reputation for corruption and oligarchic control – also deserve automatic respect and credence. By contrast, the statements that are issued by the official sources of countries that are not favored by Washington and its allies can only be allowed to consist of self-serving propaganda, no matter how accurate or perceptive they may appear to the discerning listener.

Propaganda is only something that the enemy and the enemy's media do. It is inconceivable that Western media could operate as propagandists or be involved in the active dissemination of propaganda because Western media are "free," whereas the enemy's media, even when not actually government controlled, serve merely as passive conduits for the distribution of official information. Missing from this discourse of course – quite apart from a substantial, long-established literature that incontestably exposes Western media as agencies of propaganda on all matters of great importance for the survival of elites and national interest – are the subtleties of who exactly is free in Western media to do what, when the mere establishment of a major media publishing house that enjoys direct access to powerbrokers involves enormous investments by financial interests that are often indistinguishable from the ruling elites. On the other hand, there is daily evidence, severe government constraints notwithstanding, that opposition Russian media do indeed contest the policies and ideologies of ruling elites, such as Russia's English language the *Moscow Times* and the prestige pro-Gorbachev publication, *Novaya Gazeta*. Evidence of armies of Russian Internet trolls is certainly interesting and important but needs to be placed alongside equally egregious evidence (including intelligence boasts and reports of recruiting campaigns) of similar armies of US, UK and other Western trolls acting to promote disinformation and suppress criticism online and in social media.

## Protests

When protestors take to the streets to demand a change of government and more democracy, they have the support of Western powers when those powers are hostile to the existing regime (as in Kiev in February 2014); when protestors take to the streets to demand a change of government and more democracy, they will be disowned by Western powers when those powers side with the existing regime (as in Western attitudes to the protestors in Odessa in May 2015). When protestors whom the West favors engage in strategies that break existing laws such as the invasion, occupation and even the destruction of government buildings, as happened in the Maidan protests, these are regarded as daring actions by the heroes of democracy. When protestors of whom the West disapproves engage in similar strategies, as happened in Odessa and the Donbass as protestors sought to resist intimidation of the coup regime based in Kiev, such behavior is deemed as reprehensible and hooligan. Victims of violence during protests in the lead-up to a Western-supported coup are given considerable and sympathetic coverage by Western media which also assume that the violence is the responsibility of authorities (and are reluctant or slow to acknowledge alternative evidence). Victims of violence during protests against a Western-supported coup (as happened in Odessa) are given relatively little attention or sympathy by Western media; their sufferings are minimized, blamed on the victims themselves and the identity of their persecutors is obscured or whitewashed.

## Responsibility

If Western powers have determined that a secessionist movement in a country such as Ukraine of whose regime the West approves is receiving assistance from a greater power of which the West does not approve, such as Russia, it therefore follows according to the logic of Western propaganda that the greater power, Russia (and, above all, its demonized leader Vladimir Putin), is responsible for everything that goes on in the secessionist movement and for everything that the secessionists do even though the evidence for Russian involvement is quite modest, if it exists at all, and it is clearly apparent that Russia could not possibly have such extensive and all-pervasive responsibility. Thus in the event that secessionists possibly (though this is certainly not proven at the time of writing and is itself a topic that has invited intense propaganda) shot down MH17, the Malaysian civilian airliner, then of course this must be Russia's fault, in fact Putin's fault, and was probably done deliberately, and done with arms provided by Russia (Putin) for the purpose. If secessionists, or even some rogue band of secessionists engage in attacks on (though it may equally well be regarded as a form of defense against) Ukrainian forces, in the context of Minsk I and Minsk II ceasefire treaties, then it is clearly Russia's, no, Putin's responsibility to put an immediate halt to such aggressions.

## Secession

When a region of an established country wishes to secede from that country, and the leadership of that region enjoys Western favor, then Western powers and their mainstream media will be supportive of secession. This was evident in mainstream contemplation of what might happen if the February 2104 coup failed and Kiev broke off Western Ukraine from the rest. It was evident in Western support for Kosovo's assertion of independence from Serbia. Because the leadership of Crimea and the Donbass regions did not enjoy Western favor their respective bids for secession were condemned.

If a region seeks to secede from a coup-installed regime that has the backing of Western powers, even when the regime threatens to criminalize that region's language and makes evident its disdain for the ethnicity of a large proportion of the region's population, its aspirations for autonomy are unworthy. They are deemed even less worthy if the region seeks to annex with a major neighboring power with which it is affiliated ethnically and culturally. Western support for such a region is permissible when the region in question (as in the case of Kosovo's declaration of independence from Serbia) has the support of Western powers and when the country from which it secedes is disliked by Western powers, in which case not even a referendum for secession is deemed necessary.

Much the same logic applies to movements or tendencies that are de-centering even if not to the point of secession. Centrifugal movements are good when they are de-centering from a center of which the West disapproves (e.g., Tatar resistance to incorporation of Crimea within Russia) but bad if they are de-centering from a center of which the West approves (tendencies towards federalism in Ukraine that would weaken the influence of the Kyiv government, for example).

## Suffering

The sufferings of a people who are attacked by a power allied to Western interests attract less attention, are less likely to be ennobled, than the sufferings of people who are attacked by a power that is not allied to Western interests. In the first case, suffering need not invoke intense sympathy, nor be covered in detail, because such people have brought their suffering upon themselves through the unwise choice of leaders or alliance. Similarly, killings and assassinations of figures who are opponents of regimes that the West does not like, such as Boris Nemtsov, are automatically to be cast as noble fighters of the good cause. Killings and assassinations of figures who are opponents of regimes that the West likes, such as the dozens killed among opponents to the coup regime of Kiev following February 2014, require little or no attention or sympathy.

# 11 Conclusion

In this book, I have identified the leading WMM frames and discourses of the Ukraine crisis and of its associated conflicts between the Kiev authorities and the authorities of separatist movements within Ukraine; Russia; countries of the European Union; and the USA. I have noted the Manichean forms that these discourses take. I have examined a number of these frames with a view to advancing our understanding of Western propaganda and its media conduits in the post-2001 environment of international relations. A central feature of this environment appears to be diminishing concern and respect for the Westphalian order of nation states in preference for a much more fluid if not chaotic order. Guided by my own close reading of WMM and alternative media sources, many of them also Western in origin, I have determined the principal arguments, presumptions and possible intentions of propaganda as these reflect the interests of Western powers, and considered the forms by which they are conveyed through WMM. I have several times pointed out that to deconstruct the hegemonic frames of WMM does not in itself elevate alternative or oppositional frames that are available either in the mainstream media of nations that are opposed to the Washington Consensus or may be found in media that represent resistance within the West to the narratives by which Western governments and other centers of power in the West justify their policies.

The hegemonic narrative of events that together constitute the Ukraine crisis during the period on which this book mainly focuses namely, 2014–2015, is characterized by a number of interweaving plots. Foremost amongst these is the story of a nation, Ukraine, that was once a component of the former Soviet Union – a political entity that was then and still is commonly presumed in WMM to have been a negative force in the world – but has valiantly striven to shake itself loose of the successor to the Soviet Union, the Russian Federation, in favor of forging closer links to the West and in particular to the European Union, NATO and the USA. This story of a valiant David nobly resistant through courageous struggle in the name of a people's freedom, in this case from the giant shadowy bear of the Russian Federation, has satisfyingly universal resonances. For the story to play well with WMM audiences, its authors needed to play up the virtue of the protagonists – the Ukrainian people as these appeared to the world during the Maidan protests of early 2014. It was important that their cause be seen as just. Their cause was freedom from a pro-Russian, oligarchic, corrupt and brutally oppressive regime that perversely had set its face against economic advancement, modernity and democracy. The evil of such a regime fully justified a non-compromising ouster of the existing president and his government in a scenario that its opponents would describe as a *coup d'état*. Not could the evil be destroyed with one simple blow, since in new guises it rose from the ashes to threaten the people of Ukraine through seizure of Ukrainian territory in Crimea, attempted subversion of the new order in major centers of population such as Odessa and Mariupol and, above all, in the Donetsk and Luhansk regions of the Donbass in Eastern Ukraine. Hydra-like the enemy's evil was seemingly irrepressible, made every day more apparent in Russia's active support for, if not control of, pro-Russian "rebels" in the Donbass, Russia's "responsibility" for the shooting down of civilian airliner MH17, and the authoritarian nature of the Russian regime under President Vladimir Putin who was widely held accountable for the assassination of opponents such as Boris Nemtsov and Alexander Litvinenko.

To highlight this image of an embattled but determined savior of Ukrainian nationhood, it was necessary to downplay or drown out altogether any and all discordant notes. Without returning to the deeper issues of hidden context that I discussed in earlier chapters, these notes included but were certainly not limited to the following:

1. A history of subversive US tactics to manipulate Ukrainian elections and politics to the advantage of a US-led neoliberal order;
2. The vital role played during the Maidan protests by extreme right-wing forces that openly vaunted their historical line of succession to pro-Nazi parties of World War II, and who continued to play a significant paramilitary role in the ensuing civil war;
3. Significant evidence of false-flag shootings of protestors in the Maidan protests;
4. The brutality of the governing forces in the Maidan protests was the principle evidence of brutality against the regime and even that may have been the result in part of provocation or may have been exaggerated;
5. The failure of the new authorities to complete due-process investigations of responsibility for killings in Maidan and Odessa;
6. A history of economic weakness, oligarch overreach and corruption that tainted all the principal factions in Ukraine, not simply that of the pre-Maidan incumbent;
7. A Washington-led neoliberal ambition for the Ukraine to open its borders to international corporations and finance for the acquisition of interests in privatized State assets, Ukrainian farmland and other assets;
8. The existence of a functioning democratic system under Yanukovych and his predecessors that held and would continue to have held regular, scheduled elections – thus raising the question as to whether the Maidan protestors considered that their demands could not be met through dependence on

democratic process; the implementation of "democracy" under the new regime required the abolition of political parties that the new regime disliked, including what had been the largest political party of the old regime;

9 The demographic make-up of Ukraine, especially in the South and East, did not align comfortably with the highly-centralized structure of the Ukrainian polity and with the passionate anti-Russian tenor of the new regime, and it was this more than any other factor that precipitated a bloody civil war; the so-called separatists of the Donbass actually favored more autonomy within Ukraine than they wanted alliance with Russia, and this was recognized by the West in the Minsk II agreements which called for decentralization;

10 The overthrow of Yanukovych with Western support and encouragement, assaulted previous understandings between Russia and the West that the West would not extend the reach of NATO to include countries that had previously been component States of the Soviet Union and would certainly encroach to the very borders of the Russian Federation itself;

11 Russia's presence in Crimea was legitimately sanctioned by long-established agreements between Ukraine and Russia that Russia would continue to use the port of Sebastopol for the Russian fleet; by strong pro-Russian ethnic ties; the direct threat of the new regime to the interests of ethnic Russians in Crimea;

12 Evidence of Russian responsibility for the shooting down of MH17 – Dutch Safety Board and Joint International Team reports notwithstanding – continues to be contested. Even were Russia or ethnic Russian separatists in Eastern Ukraine actually responsible for the tragedy, this would not detract from possible criminal negligence of the Ukraine authorities and of international airlines in continuing to keep open and fly over space where it was known that Buk missile launchers were positioned and where several military planes had already been shot down;

13 For every claim of Russian intervention, direct or indirect, in the Donbass there was evidence of Western provision of lethal and non-lethal assistance to the Ukrainian forces and paramilitaries;

14 For every claim of Russian or ethnic Russian human rights abuse, curtailment of free expression in the Donbass and Crimea, and propaganda there were equivalent claims that could be made of Ukrainian authorities and their strongest allies. On the Western side, these were aggravated by evidence of a clear campaign against Russia's economy through sanctions, possibly through the manipulation of oil prices, and attempts to destabilize Russian plans for the extension of oil and gas pipelines. In addition, Western wars, fought on false pretext and involving the invasion and occupation of sovereign territories, together with Western support for the aggression of its allies in countries such as Syria and Yemen, and with the increasing Western dependence on frequently imprecise and indiscriminate methods of assassination such as the use of drones, not to mention instances of suspected State homicide closer to home, easily counteracted and significantly outpaced claims of Russian responsibility for the assassination of State opponents.

Determined readers could certainly find through the proliferation of available publications and broadcasts, both online and off-line, evidence or accounts that challenged the hegemonic narratives of events and the discourses as to what the Ukraine crisis signified. Propaganda for the most part functions on the assumption that very few media consumers are especially determined searchers for information, least of all on matters of relatively obscure foreign policy. The principal sources of current affairs information on the Internet continue to be dominated by legacy mainstream media publishers and broadcasters and their online progeny. Their voices shout the loudest and are the most "authoritative" (because they are closest to the main sources of political and economic power). They are also likely to be the most influential for most of the time – even to the point of providing frames that are adopted in whole or in part by legions of future journalists, policy makers and academics eager to believe that the basic building blocks of "what really happened" and "what did it really mean" are securely in place and provide firm support for more elaborate constructions.

# 12 Postscript

WMM vilification of the Russian President Vladimir Putin approached absurdity with coverage of the "Panama Papers" tax avoidance scandal in April 2016 that they spearheaded with false insinuations of Putin's involvement. Putin's name was not mentioned in records of tax evasion from the law firm Mossack Fonseca and released by the International Consortium of Investigative Journalists (ICIJ) (Harding 2016). The significance of Putin's possible ties to Russians who *were* mentioned was sketchy in the extreme and much less important than the involvement of US, British, Chinese and other oligarchs globally.

In short, the Ukraine crisis had helped reignite Cold War tensions, accentuating fears of nuclear confrontation, whether planned or accidental. A former NATO Deputy Supreme Allied Commander predicted that Russia would attack NATO in 2017 (Robinson 2016). In May, NATO deployed the Aegis Ashore ballistic missile defense system in Romania and planned a further deployment in Poland in 2018. Russia protested that such systems are anything but "defensive" since they qualify as launch areas for US missiles. NATO forces engaged in large-scale military exercises close to the Russian border in Poland and in Ukraine in June. Meanwhile, Russia tested prototypes of the S-500 Prometey air and missile defense system capable of destroying ICBMs, hypersonic cruise missiles and up to ten ballistic missile warheads before these should re-enter the atmosphere (Escobar 2016).

Russia's economy was damaged significantly by EU and US sanctions (which, together with Russian sanctions, also negatively impacted European and Ukrainian economies). Yet, increased self-sufficiency boosted the prospects for longer-term recovery with less debt. Russia benefited from a modest recovery of global oil prices. Ukrainian, US and EU obstacles to Russian ambitions to supply more oil and gas to Europe and beyond were partly circumvented with the help of German collaboration and improved relations with Turkey.

Internal conflict erupted within the EU and NATO over whether sanctions against Russia should be maintained (they were renewed until the end of 2016), and whether Russia constituted any kind of real threat to Europe. The United Kingdom's dramatic "Brexit" vote in June augured the possible disintegration of the EU, opening the door to better relations between Russia and Europe at the expense of US maneuvers for Eurasian domination. Russia and China consolidated political, industrial and trading ties.

Fighting between Ukrainian and separatist forces in the Donbass flared repetitively, with each side accusing the other of numerous violations. WMM criticized the Supreme Court of Crimea for labeling the Tatar Mejlis as a terrorist organization, now banned, but underplayed its leaders' engagement in economic terrorism against Crimea. Russia successfully established direct sources of electricity supply from the mainland. A Gfk poll from February 2015, sponsored by a pro-Ukrainian group in Canada, revealed that 93 percent of Crimeans still endorsed the 2014 annexation referendum (Kosaretsky 2015).

Ukraine prosecutors achieved little formal progress in their investigations of either the Maidan or Odessa massacres. Some Ukrainians suspected authorities were protecting pro-Yanukovych agents. Kiev's investigations unhelpfully treated the killings of police separately from the killings of civilians and did not admit of the possibility of a false-flag operation. By contrast, a detailed investigation by Ivan Katchanovski of the University of Ottawa offered significant evidence that killings of protestors were organized by far right paramilitary groups (Katchanovski 2015c; *Sputnik News* 2016).

A pro-Kiev documentary, *Winter on Fire: Ukraine's Fight for Freedom*, edited out the role of neo-Nazi paramilitaries (Afineevsky 2015). These remained central to French documentarist Paul Moreira's *Ukraine: The Masks of the Revolution* (Moreira 2016) and Leeor Kaufman's *Maidan: Tonight and Tomorrow* (Kaufman 2016). Oliver Stone's *Ukraine on Fire* included material on the machinations of NGOs. Ukrainian authorities protested the BBC's *The Conspiracy Files: Who Shot Down MH17* (Rudin 2016), which allowed for the possibility of an air-to-air missile attack. The Ukrainian Institute of National Memory sought to retell the narrative of Ukraine's history by amplifying – according to its critics – Soviet crimes, glorifying Ukrainian nationalist fighters, and downplaying evidence of ethnic cleansing of Poles and Jews (Cohen 2016).

President Poroshenko faced continuing pressure from right-wing militia who vowed to oust him if he allowed elections in the Donbass – as the Minsk II agreement would require. The Minsk process should have been completed by the end of 2015. Concern about systemic corruption was barely relieved when Rada replaced Prime Minister Yatsenyuk with Volodymyr Groysman in April. In 2015, Ukraine ranked 130 in the list of the 168 countries in the Corruption Perception Index (Transparency International 2016). Yatsenyuk's departure triggered the IMF's resumption of lending that had stalled after August 2015.

New lending went to a weak economy whose inflation had peaked at 43 percent in 2015. Prices increased by 79 percent, 2013–2015. The public debt-to-GDP ratio was 79 percent at the end of 2015. GDP declined by 10.5 percent that year. Forty-two percent of Ukraine's economic activity remained unaccounted for. In the first year of its association with the EU, Ukrainian exports to the EU actually fell by 30 percent. Ukraine became ever more an agricultural economy. Much extraction and manufacturing activity was concentrated in the war-torn east. High-tech sectors had all been integrated with Russian (various data sources).

The Dutch-led Joint International Criminal Investigation Team released a progress report on its MH17 investigation in June (JIT 2016). This emphasized the challenges of training in appropriate investigative methods and the value of a consensual approach. Participating governments – the Netherlands, Belgium, Australia, Malaysia and Ukraine (but not Russia) – had the right to block the release of information to the public. The JIT predicted a final report before the end of 2016, but said this would not be open for public examination. Working in Kiev, team members reported positively on collaboration – both internal and with Ukrainian authorities. These included the SBU, Ukraine's intelligence agency – a significant source of evidence (of dubious quality, many critics of the SBU would argue).

The consistently critical voice of the investigative reporter Robert Parry identified these disclosures as indicative of potential conflicts of interest. He reminded readers of earlier investigations that concluded that all BUKs in Eastern territory on June 17, 2014 were under the control of Ukrainian military (hence explaining the focus of some alternative investigators on a supposed BUK they allege was driven into the Donbass from Russia) and that the rebels had only "MANPADS" incapable of firing up to 33,000 feet. Because front lines were shifting rapidly on that day, it was hard to know where government and rebel lines actually lay.

Parry alluded to CIA sources that kept open the possibility of MH17 having been targeted by a "rogue" Ukrainian unit which mistook MH17 for the official plane flying Vladimir Putin back to Russia from Latin America that day. Parry also remained open to the possibility of an air-to-air missile, in view of seven eyewitness statements in Ukraine who reported seeing a warplane just before MH17 was hit. The US State Department persisted in its refusal to make public its own surveillance evidence. The chief investigator of the Dutch National Prosecutor's Office claimed early in 2016 that it possessed no video footage of the missile launch – due to "cloud cover" – and had no satellite images. The Ukrainian primary radar systems were not functioning on July 17, supposedly for reasons of malfunction and maintenance. The Ukraine's secondary radar system would not have tracked military aircraft since these do not identify themselves with transponders. Russia claimed that it had detected a possible warplane but its primary radar data was not saved, because Eastern Ukraine is outside its jurisdiction. Russia claimed to have offered its visual screen images to the JIT but that these were discounted by the investigators (Parry 2016c, 2016d).

# Bibliography


Afineevsky, E. (Dir) (2015) *Winter on Fire: Ukraine's Fight for Freedom*. Documentary film: Netflix (producer, distribution).

AFP (2014) "Kiev Says Captured 10 Russian Paratroopers in East Ukraine." *Global Post*, August 25, at: www.globalpost.com/dispatch/news/afp/140825/kiev-says-captured-10-russian-paratroopers-east-ukraine (accessed August 29, 2014).

AFP (2015a) "Putin Accuses US over Ukraine Coup." *Daily Nation*, March 17, at: www.nation.co.ke/news/world/Putin-accuses-US-over-Ukraine-coup/-/1068/2657280/-/ngwlb4/- (accessed March 28, 2015).

AFP (2015b) "Ukraine Offers Huge State Firms to Foreign Investors." *AFP*, July 9, at: www.yahoo.com/news/ukraine-offers-huge-state-firms-foreign-investors-153701811.html?ref=gs (accessed July 21, 2015).

AFP (2015c) "Ukraine Halts Russian Gas Purchases, Closes its Airspace." *Channelnews-agency.com*, November 26, at: www.channelnewsasia.com/news/world/ukraine-bans-russia-s/2292934.html (accessed December 24, 2015).

Ahmed, N.M. (2006) *The London Bombings: An Independent Inquiry*. New York: Overlook Press.

Amos, H. (2015) "Ex-U.K. Ambassador a Lonely Friend of Russia." *Moscow Times*, October 5, at: www.themoscowtimes.com/news/article/ex-uk-ambassador-a-lonely-friend-of-russia/537205.html (accessed December 6, 2015).

Andersen, J. (2015a) "Yatsenyuk Rebuffs Criticism Over Gas Price Hikes, Takes Aim at Kolomoisky." *Kiev Post*, November 20, at: www.kyivpost.com/content/ukraine/yatsenyuk-rebuffs-criticism-over-gas-price-hikes-takes-aim-at-kolomoisky-398225.html (accessed December 8, 2015).

Andersen, J. (2015b) "Frustrations High at EuroMaidan Killings Hearing as Violence Erupts." *Kyiv Post*, November 20, at: www.kyivpost.com/content/ukraine/frustrations-high-at-euromaidan-killings-hearing-as-violence-erupts-402499.html (accessed December 8, 2015).

Annis, R. (2015) "State of Emergency in Crimea: Right-Wing Extremists in Ukraine Blow Up Electricity Lines To the Peninsula." *Counterpunch*, November 25, at: www.counterpunch.org/2015/11/25/state-of-emergency-in-crimea (accessed December 3, 2015).

Anon (2014) "HRW: Civilian Death Toll in E. Ukraine Rising Due to 'indiscriminate and Unlawful' Shelling." *RT*, September 2, at: http://rt.com/news/184400-hrw-ukraine-indiscriminate-shelling (accessed September 6, 2014).

Antonova, M. (2015) "They Came to Fight for Ukraine. Now They're Stuck in No Man's Land." *Foreign Policy*, October 19, at: http://foreignpolicy.com/2015/10/19/ukraines-abandoned-soldiers-russian-belarusian-volunteers (accessed December 6, 2015).

Armistead, E.L. (2004) *Information Operations: Warfare and the Hard Reality of Soft Power*. Washington, DC: Brassey's Inc.

Baker, P. (2014) "Sanctions Revive Search for Secret Putin Fortune." *New York Times*, April 27, at: www.nytimes.com/2014/04/27/world/sanctions-revive-search-for-secret-putin-fortune.html?_r=0 (accessed July 22, 2014).

Baker, P. and Erlanger, S. (2015) "Russia Uses Money and Ideology to Fight Western Sanctions." *New York Times*, June 7, at: www.nytimes.com/2015/06/08/world/europe/russia-fights-wests-ukraine-sanctions-with-aid-and-ideology.html?_r=0 (accessed June 29, 2016).

Bauerova, L. and Carroll, J. (2014) "Chevron Shale Exit Shreds Ukraine's Hope of Energy Independence." *Bloomberg*, December 17, at: www.bloomberg.com/news/2014-12-17/chevron-shale-exit-shreds-ukraine-s-hope-of-energy-independence.html (accessed January 2, 2015).

BBC (2015) "Russian Profile – Media." *BBC News*, February 25, at: www.bbc.co.uk/news/world-europe-17840134 (accessed June 20, 2016).

Becker, J. and Myers, S. (2015) "Russian Groups Crowdfund the War in Ukraine." *New York Times*, June 11, at: www.nytimes.com/2015/06/12/world/europe/russian-groups-crowdfund-the-war-in-ukraine.html?ref=topics&_r=0 (accessed July 14, 2015).

Bellant, P. (1991) *Old Nazis, the New Right, and the Republican Party: Domestic Fascist Networks and their Effect on US Cold War Politics*. 3rd edn. Cambridge, MA: South End Press.

Bennett, W.L., Lawrence, R.G. and Livingston, S. (2007) *When the Press Fails: Political Power and the News Media from Iraq to Katrina*. Chicago, IL: University of Chicago Press.

Bidder, B. (2014) "Dossier of Dubiousness: Did Putin's Man in Crimea have Mafia Ties?" *Spiegel Online*, March 25, at: www.spiegel.de/international/world/dossier-suggests-crimea-prime-minister-aksyonov-had-crime-ties-a-960644.html (accessed June 14, 2015).

Blum, W. (2014) "The New Cold War: The Russians are Coming.... Again ... and They're Still Ten Feet Tall." *Global Research*, May 10, at: www.globalresearch.ca/the-new-cold-war-the-russians-are-coming-again-and-theyre-still-ten-feet-tall/5381594 (accessed July 21, 2014).

Blum, W. (2015) "Mysterious Deaths in Ukraine." *Consortiumnews.com*, April 3, at: https://consortiumnews.com/2015/04/03/mysterious-deaths-in-ukraine (accessed April 22, 2015).

Blunden, B. (2014) "Pay No Attention to That Man Behind the Curtain: The Media ignores the CIA in Ukraine." *Democracy Web*, August 4, at: http://democracyweb.com/?p=24606 (accessed August 6, 2014).

Boardman, W. (2014) "US Government Backs Some War Crimes, Not Others." *Reader Supported News*, August 23, at: http://readersupportednews.org/opinion2/277-75/24934-focus-us-government-backs-some-war-crimes-not-others (accessed August 10, 2014).

Boardman, W. (2015a) "French Rap US and NATO for Playing Russian Roulette with Ukraine." *Reader Supported News*, June 3, at: http://readersupportednews.org/opinion2/277-75/30528-focus-french-rap-us-and-nato-for-playing-russian-roulette-with-ukraine (accessed June 7, 2015).

Boardman, W. (2015b) "Ukraine Peace Move Sets Off Violent Rightist Riot in Kiev." *Reader Supported News*, September 15, at: http://readersupportednews.org/opinion2/277-75/32278-focus-ukraine-peace-move-sets-off-violent-rightist-riot-in-kiev (accessed December 24, 2015).

Bodner, M. (2015a) "5 Heavy Weapons in Ukraine's Rebel War That May Have Come from Russia." *Moscow Times*, February 6, at: www.themoscowtimes.com/business/article/5-heavy-weapons-in-ukraine-s-rebel-war-that-may-have-come-from-russia/515557.html (accessed January 2, 2015).

Bodner, M. (2015b) "Russia Defense Ministry Doctored MH17 Satellite Images." *Moscow Times*, June 1, at: www.themoscowtimes.com/news/article/report-russias-defense-ministry-doctored-mh17-satellite-images/522855.html (accessed June 7, 2015).

Bodner, M. (2015c) "Kremlin Threatens Response to US Nuclear Bomb Deployment in Germany." *Moscow Times*, September 23, at: www.themoscowtimes.com/business/article/kremlin-threatens-response-to-us-nuclear-bomb-deployment-in-germany/535106.html (accessed December 8, 2015).

Boiko, A. (2015) "Dutch Experts Detect No Russian Activities in Rostov Region." *Sputnik News*, February 20, at: http://sputniknews.com/military/20150220/1018509984.html (accessed March 2, 2015).

Borger, J. and Higgins, E. (2015) "How Blogger Brown Moses used Open-Source Information to Analyze Missile Attacks on Ukraine." *Guardian*, February 17, at: www.theguardian.com/world/video/2015/feb/17/brown-moses-cross-border-artillery-analysis-video (accessed February 19, 2015).

Borozna, A. (2015) "The Evolution of the West's Negative Narrative about Russia." *Opednews.com*, August 27, at: www.opednews.com/articles/The-evolution-of-negative-by-Angela-Borozna-Media-Distortion_Media-Lies_Putin_Russia-150827-265.html (accessed August 31, 2015).

Boyd-Barrett, O. (2003) "Doubt Foreclosed: U.S. Mainstream Media and the Attacks of 9–11." In N. Chitty, R. Rush and M. Semati (eds.), *Studies in Terrorism*. Penang: Journal of International Communication/ Southbound Press, pp. 35–54.

Boyd-Barrett, O. (2015) *Media Imperialism*. London and Thousand Oaks, CA: Sage.

Boyd-Barrett, O., Herrera, D. and Baumann, J.A. (2011) *Hollywood and the CIA: Cinema, Defense and Subversion*. London: Routledge.

Britic (2015) "Confession of a CIA Agent: They Gave us Millions to Dismember Yugoslavia." *Global Research*, November 29, at: www.globalresearch.ca/confession-of-a-cia-agent-they-gave-us-millions-to-dismember-yugoslavia/5492008 (accessed January 13, 2016).

Bullough, O. (2014) "Stop Forcing Ukraine into a Narrative of Moscow Versus Washington." *Guardian*, May 19.

Carden, J. (2015) "Ukraine Exposed: Kiev's Authoritarianism." *The National Interest*, January 30, at: http://nationalinterest.org/feature/ukraine-exposed-kievs-authoritarianism-12151 (accessed February 7, 2015).

Carden, J. (2016) "Congress Has Removed a Ban on Funding Neo-Nazis From Its Year-End Spending Bill." *The Nation*, January 14, at: www.thenation.com/article/congress-has-removed-a-ban-on-funding-neo-nazis-from-its-year-end-spending-bill (accessed January 19, 2016).

Cartalucci, T. (2014a) "Dutch MH17 Investigation Omits US 'Intel.' Fabrications and Omissions Supportive of US-NATO Agenda Directed against Russia." *Global Research*, September 19, at: www.globalresearch.ca/dutch-mh17-investigation-omits-us-intel-fabrications-and-omissions-supportive-of-us-nato-agenda-directed-against-russia/5402970 (accessed December 1, 2014).

Cartalucci, T. (2014b) "The 'Impending' Russian Madian." *New Eastern Outlook*, December 28: at: http://journal-neo.org/2014/12/20/the-impending-russian-maidan (accessed March 1, 2015).

Cathcart, W. and Epstein, J. (2015) "How Many Neo-Nazis is the US Backing in Ukraine?" *The Daily Beast*, June 9, at: www.thedailybeast.com/articles/2015/06/09/how-many-neo-nazis-is-the-u-s-backing-in-ukraine.html (accessed July 20, 2015).

Chivers, C.J. and Reevel, P. (2014) "Russia Moves Swiftly to Stifle Dissent Ahead of Secession Vote." *New York Times*, March 15, at: www.nytimes.com/2014/03/15/world/europe/pressure-and-intimidation-sweep-crimea-ahead-of-secession-vote.html?_r=0 (accessed May 6, 2015).

Chivers, C.J. and Sneider, N. (2014) "Behind the Masks in Ukraine, Many Faces of Rebellion." *New York Times*, May 4, at: www.nytimes.com/2014/05/04/world/europe/behind-the-masks-in-ukraine-many-faces-of-rebellion.html?_r=0 (accessed June 26, 2014).

Cloughley, B. (2015) "NATO is Building Up for War: The Rising Swell of Anti-Russian Propaganda." *Counterpunch*, April 3–5, at: www.counterpunch.org/2015/04/03/nato-is-building-up-for-war (accessed April 27, 2015).

Coalson, R. (2015) "Crimean Tatars Push Kyiv's Hand On Recovering Crimea." *RFE/RL*, November 24, at: www.rferl.org/content/crimean-tatars-push-kyiv-hand-on-recovering-crimea/27384563.html (accessed December 6, 2015).

Cockburn, A. (2015) "Undelivered Goods: How $1.8 Billion in Aid to Ukraine was Funneled to the Outposts of the International Finance Galaxy." *US-Russia.org*, August 13, at: http://us-russia.org/3476-undelivered-goods-how-18-billion-in-aid-to-ukraine-was-funneled-to-the-outposts-of-the-international-finance-galaxy.html (accessed August 30, 2015).

Cohen, J. (2016) "The Historian Whitewashing Ukraine's Past." *Foreign Policy*, May 2, at: http://foreignpolicy.com/2016/05/02/the-historian-whitewashing-ukraines-past-volodymyr-viatrovych (accessed June 29, 2016).

Collins, M. (2014a) "Eyewitness Evidence of Machine Gun Fire on MH17." *The Peoples Voice*, August 5, at: www.thepeoplesvoice.org/TPV3/Voices.php/2014/08/05/eyewitness-evidence-of-machine-gun-fire (accessed December 27, 2014).

Collins, M. (2014b) "Ukraine Under Water to IMF." *OpEd News*, December 13, at: www.opednews.com/articles/Ukraine-Under-Water-to-IMF-by-Michael-Collins-European-Union_International-Monetary-Fund-IMF_Russia_Ukraine-141213-621.html (accessed January 19, 2015).

Collins, M. (2014c) "Shooting Down MH17 – BUK 312 Story False says Ukraine Crew Member." *Global Research*, December 17, at: www.globalresearch.ca/shooting-down-mh-17-buk-312-story-false-says-ukraine-crew-member/5420405 (accessed on December 26, 2014).

Cook, J. (2014) "More Guardian 'Brainwashing' on Putin." *Jonathan-Cook.net*, March 24, at: www.jonathan-cook.net/blog/2014-03-24/more-guardian-brainwashing-on-putin (accessed June 24, 2014).

Creel, G. (2012) *How We Advertised America: The First Telling of the Amazing Story of the Committee on Public Information That Carried the Gospel of Americanism to Every Corner of the Globe*. Ulan Press.

Cromwell, D. and Edwards, D. (2014) "The 'Professorial President' and the 'Small, Strutting Hard Man.'" *Media Lens*, March 10, at: www.medialens.org/index.php/alerts/alert-archive/2014/757-the-professorial-president-and-the-small-strutting-hard-man.html (accessed May 28, 2016).

Cunningham, F. (2014) "CIA Mouthpiece Lets Slip on Russian 'Aggression.'" *Information Clearing House*, October 30, at: www.informationclearinghouse.info/article40099.htm (accessed January 2, 2015).

Cunningham, F. (2015a) "NATO Finds Arab Backdoor to Arm Kiev." *Strategic Culture Foundation*, February 25, at: www.strategic-culture.org/pview/2015/02/25/nato-finds-arab-backdoor-to-arm-kiev.html (accessed July 23, 2015).

Cunningham, F. (2015b) "Entire Western Media is a Troll Army." *Sputnik News*, April 20, at: http://sputniknews.com/columnists/20150420/1021135202.html (accessed April 22, 2015).

Cunningham, F. (2015c) "Washington Choreographing All-Out War with Russia?" *Strategic Culture Foundation*, April 26, at: http://m.strategic-culture.org/news/2015/04/26/washington-choreographing-all-out-war-with-russia.html (accessed April 27, 2015).

Cunningham, F. (2015d) "Trapping Russia to Prove a Negative." *Sputnik News*, June 11, at: http://sputniknews.com/columnists/20150610/1023184358.html (accessed July 22, 2015).

Curtis, A. (2002) *The Century of the Self*. Adam Curtis Films, at: http://adamcurtisfilms.blogspot.com (accessed January 2, 2016).

Curtis, A. (2004) *The Power of Nightmares*. Adam Curtis Films, at: http://adamcurtisfilms.blogspot.com (accessed January 2, 2016).

C.W. (2014) "Why the Ukrainian Economy is in Deep Trouble." *Economist*, December 1, at: www.economist.com/blogs/economist-explains/2014/12/economist-explains (accessed January 19, 2015).

Czuperski, M., Herbst J., Higgins E., Polyakova A. and Wilson, D. (2015) *Hiding in Plain Sight: Putin's War in Ukraine*. Atlantic Council, May 27, at: www.atlanticcouncil.org/publications/reports/hiding-in-plain-sight-putin-s-war-in-ukraine-and-boris-nemtsov-s-putin-war (accessed June 7, 2015).

Dapkus, L. and Keyton, D. (2015) "NATO Opens Military Center in Lithuania Amid Ukraine Crisis." *Associated Press*, September 3, at: www.voanews.com/content/ap-lithuania-nato-center-ukraine-russia/2943692.html (accessed December 21, 2015).

Dejevsky, M. (2015) "The More the West sees the Ukraine Crisis as Another Cold War the Worse It will Get." *Information Clearing House*, February 9, at: www.informationclearinghouse.info/article40945.htm (accessed February 9, 2015).

Deutsch, J. (2015) "Russia, Ukraine and US Hegemony." *Counterpunch*, July 16, at: www.counterpunch.org/2015/07/09/russia-ukraine-and-u-s-hegemony (accessed July 23, 2015).

Dickinson R., Linne, O. and Harindranath, R. (1998) *Approaches to Audiences*. London: Edward Arnold.

Ditz, J. (2014a) "Despite Hysteria on Malaysia Airlines MH17, Little Solid Evidence 'Smoking Gun' Evidence from Ukraine Spies of Dubious Origin." *Antiwar.com*, July 18, at: http://news.antiwar.com/2014/07/18/despite-hysteria-on-malaysia-airlines-mh17-little-solid-evidence (accessed August 6, 2014).

Ditz, J. (2014b) "State Dept Contradicts Intel Community, Says Putin Directly to Blame for MH17 Claims Even More Secret Evidence Proves Putin's Involvement." *Antiwar.com*, July 23, at: http://news.antiwar.com/2014/07/23/state-dept-contradicts-intel-community-says-putin-directly-to-blame-for-mh17 (accessed August 11, 2014).

Ditz, J. (2014c) "US invents Reports of Russia attacking Ukraine Bases: No Reports out of Ukraine on Any Such Incidents." *AntiWar.com*, July 24, at: http://news.antiwar.com/2014/07/24/us-invents-reports-of-russia-attacking-ukraine-bases (accessed August 7 2014).

Ditz, J. (2015a) "German Officials: Claims of Russian Invasion of Ukraine are American Propaganda." *Mint Press*, March 13, at: www.mintpressnews.com/german-officials-claims-of-russian-invasion-of-ukraine-are-american-propaganda/203338 (accessed March 15, 2015).

Ditz, J. (2015b) "US to Preposition Tanks, Artillery Along Russian Border: Defense Secretary Claims Move Russia's Fault." *Antiwar.com*, June 23, at: http://news.antiwar.com/2015/06/23/us-to-preposition-tanks-artillery-along-russian-border (accessed July 29, 2015).

Ditz, J. (2015c) "Ukraine President Orders Disarming of Far-Right Factions, Says No Political Faction Should Have Armed Cells." *Antiwar.com*, July 13, at: http://news.antiwar.com/2015/07/13/ukraine-president-orders-disarming-of-far-right-factions (accessed July 16, 2015).

Djurica, M. (2016) "Europe Lays Out Conditions for Russia's South Stream Return." *RT.com*, February 10, at: www.rt.com/business/331995-south-stream-renewal-conditions (accessed July 4, 2016).

Doctorow, G. (2016) "Russian Documentary Film aims to Avert Nuclear War – Western Media Still Busy Demonizing Putin." *Russia Insider*, January 2, at: http://russia-insider.com/en/politics/russian-documentary-aims-avert-nuclear-war-western-media-still-busy-demonizing-putin (accessed January 7, 2016).

Dolgov, A. (2015) "Nemtsov Allies Plan to Publish Report on Russian Soldiers in Ukraine." *Moscow Times*, March 31, at: www.themoscowtimes.com/news/article/nemtsov-allies-plan-to-publish-report-on-russian-soldiers-in-ukraine/518314.html (accessed January 1, 2016).

Dorrell, F. (2005) *What I've Learned about US Foreign Policy*. DVD-ROM, at: www.addictedtowar.com/dorrel.html (accessed January 1, 2016).

Dreier, C. (2014) "German Chancellor Angela Merkel visits Kiev." *World Socialist Web Site*, August 23, at: www.wsws.org/en/articles/2014/08/26/ukra-a26.html (accessed August 27, 2014).

Dunkerley, W. (2016) "Six Reasons You Can't Take the Litvinenko Report Seriously." *Guardian*, February 5, at: www.theguardian.com/profile/william-dunkerley (accessed February 8, 2016).

Durden, T. (2014a) "Did Ukraine Attack its Own Tanks? White House 'Can't Confirm Russian Convoy was Destroyed by Kiev'." *Zero Hedge*, August 15, at: https://newsana.com/breaking-news/story/news-2014-08-15-did-ukraine-attack-its-own-tanks-white-house-cant-confirm-russian-convoy-was-destroy (accessed August 19, 2014).

Durden, T. (2014b) "Company in Which US Vice President Joe Biden's Son is Director Prepares to Drill Shale Gas in Eastern Ukraine." *Global Research*, November 21, at: www.globalresearch.ca/company-in-which-us-vice-president-joe-bidens-son-is-director-prepares-to-drill-shale-gas-in-east-ukraine/5393403 (accessed June 20, 2016).

Durden, T. (2015a) "Europe Fractures: France Pivots to Putin, Cyprus Offers Moscow Military Base, Germany-US Splinter on Ukraine." *Information Clearing House*, February 8, at: www.informationclearinghouse.info/article40939.htm (accessed February 9, 2015).

Durden, T. (2015b) "Ukraine Enters Hyperinflation: Currency Trading Halted, 'Soon We Will Walk Around with Suitcases for Cash'." *Zero Hedge*, February 25, at: www.zerohedge.com/news/2015-02-25/ukraine-enters-hyperinflation-currency-trading-halted-soon-we-will-wwalk-around-suit (accessed March 6, 2015).

DW (2015a) "No Russian Troops in Ukraine, Putin Says." *Deutsche Welle*, April 16, at: www.dw.de/no-russian-troops-in-ukraine-putin-says/a-18387284 (accessed April 27, 2015).

DW (2015b) "German Exports to Russia take a Tumble." *Deutsche Welle*, May 21, at: www.dw.com/en/german-exports-to-russia-take-a-tumble/a-18464606 (accessed July 4, 2016).

Eade, K.G. (2015) "GMOs Ukraine Connection." *OpEdNews.com*, May 20, at: www.opednews.com/articles/GMO-s-Ukrainian-Connection-by-Kenneth-G-Eade-Agribusiness_Agriculture_Biotechnology_Food-150522-851.html (accessed June 3, 2015).

Edwards, D. (2014) "'Hard Clay' – Remaking Afghanistan in 'Our' Image." *MediaLens*, April 28, at: www.medialens.org/index.php/alerts/alert-archive/2014/762-hard-clay-remaking-afghanistan-in-our-image.html (accessed June 24, 2014).

Eliason, G. (2014) "The Nazis Even Hitler was Afraid Of." *OpEdNews.com*, March 14, at: www.opednews.com/articles/The-Nazis-Even-Hitler-was-by-George-Eliason-Exile_Genocide_Hitler-Brownshirts_Hitler-Youth-140316-321.html (accessed June 26, 2014).

Eliason, G. (2015a) "Mariupol Attack is a War Crime: The Evidence Points to Kiev." *Global Research*, January 29, at: www.globalresearch.ca/mariupol-attack-is-a-war-crime-the-evidence-points-to-kiev/5428194 (accessed March 17, 2015).

Eliason, G. (2015b) "Can the Ukrainian Government target American Journalists in America?" *OpEd News*, March 19, at: www.opednews.com/articles/Can-the-Ukrainian-Governme-by-George-Eliason-American-Terror_Americans-Killed_Government-Bullying_Government-Crime-150319-980.html (accessed March 26, 2015).

Engdahl, W. (2011) "The Real Crime of M. Khodorkovsky." *Voltairenet.org*, January 5, at: www.voltairenet.org/article168007.html (accessed January 19, 2016).

Engdahl, W. (2015) "US's Saudi Oil Deal from Win–Win to Mega-Loose." *Landdestroyer.blogspot.com*, August 10, at: http://landdestroyer.blogspot.com/2015/08/uss-saudi-oil-deal-from-win-win-to-mega.html (accessed June 20, 2016).

Entman, R. (2003) *Projections of Power: Framing News, Public Opinion and US Foreign Policy*. Chicago, IL: University of Chicago Press.

Escobar, P. (2007) *Globalistan: How the Globalized World is Dissolving into Liquid War*. Ann Arbor, MI: Nimble Books.

Escobar, P. (2014a) "Ukraine: The Waiting Game." *Asia Times*, May 13, at: www.atimes.com/atimes/Central_Asia/CEN-01-130514.html (accessed June 20, 2016).

Escobar, P. (2014b) "The Tragedy of MH17: A Chessboard Drenched in Blood." *Asia Times*, July 23, at: www.atimes.com/atimes/Central_Asia/CEN-01-230714.html (accessed August 10, 2014).

Escobar, P. (2014c) "From Minsk to Wales, Germany is the Key." *RT*, August 28, at: http://rt.com/op-edge/183328-minsk-wales-germany-key (accessed September 6, 2014).

Escobar, P. (2015a) "Eurasian Emporium or Nuclear War?" *Asia Times*, April 6, at: http://atimes.com/2015/04/eurasian-emporium-or-nuclear-war-pepe-escobar/ (accessed June 20, 2016).

Escobar, P. (2015b) "Why the US is Finally Talking to Russia." *Information Clearing House*, May 20, at: www.informationclearinghouse.info/article41919.htm (accessed June 3, 2015).

Escobar, P. (2015c) "Reshuffling Eurasia's Energy Deck – Iran, China and Pipelineistan." *Counterpunch.org*, August 4, at: www.counterpunch.org/2015/08/04/reshuffling-eurasias-energy-deck-iran-china-and-pipelineistan (accessed August 31, 2015).

Escobar, P. (2016) "Beware What You Wish For: Russia is Ready for War." *Mint Press News*, May 23, at: www.mintpressnews.com/beware-wish-russia-ready-war/216625 (accessed June 29, 2016).

Faiola, A. (2014) "In Ukraine, a Crisis of Bullets and Economics." *Washington Post*, April 16, at: www.washingtonpost.com/world/2014/04/16/4d36b1b6-c532-11e3-b574-f8748871856a_story.html (accessed June 26, 2014).

Fedets, I. (2015) "Oligarchs on the Airwaves." *Foreign Policy*, November 11, at: http://foreignpolicy.com/2015/11/11/oligarchs-on-the-airwaves-ukraine-media (accessed December 20, 2015).

Ferris-Rotman, A. (2015) "Life in Ukraine's Buffer Zone: The Front Line of the War Divides Towns, Villages and Even Families." *Politico*, December 6, at: www.politico.eu/article/life-in-ukraines-buffer-zone (accessed January 7, 2016).

Fitzpatrick, C. (2014a) "Evidence of Separatists' Possession of Buk System Before Downing of MH17." *The Interpreter*, July 27, at: www.interpretermag.com/evidence-of-separatists-possession-of-buk-system-before-downing-of-mh17 (accessed January 21, 2016).

Fitzpatrick, C. (2014b) "Novaya Gazeta Editor: Boroday Called Moscow Press About Downing of Civilian Airliner." *The Interpreter*, August 13, at: www.interpretermag.com/novaya-gazeta-editor-boroday-called-moscow-press-about-downing-of-civilian-airliner (accessed January 21, 2016).

Forbidden Knowledge (2014) "MH17 Witness Passes Lie Detector Test." *Forbidden Knowledge*, December 23, at: www.ForbiddenKnowledgeTV.com/page/26916.html (accessed December 26, 2014).

Ford, G. (2014) "Obama's 'Hot War.'" *Common Dreams*, July 23, at: www.commondreams.org/views/2014/07/23/obamas-hot-war (accessed August 10, 2014).

Franklin B., Hamer, M., Hanna, M. & Kinsey, M. (2005) *Key Concepts in Journalism Studies*. Thousand Oaks, CA: Sage.

Friel, H. (2014) "Ukraine and International Law: The Double Standard at the *New York Times* Editorial Page." *Common Dreams*, at: www.commondreams.org/views/2014/05/19/ukraine-and-international-law-double-standard-new-york-times-editorial-page (accessed August 11, 2014).

Giraldi, P. (2015) "MH17: The Blaming Putin Game Goes On. Who Shot Down MH-17? Somebody Knows." *The Unz Review*, July 14, at: www.unz.com/article/mh17-the-blaming-putin-game-goes-on (accessed July 23, 2015).

Glaser, B. G. and Strauss, A. L. (1967). *The Discovery of Grounded Theory: Strategies for Qualitative Research*. Chicago, IL: Aldine Publishing Company.

Glynn, C., Herbst, S.Lindeman, M., O'Keefe, J., & Shapiro, R. (2015) *Public Opinion*. Boulder, CO: Westview Press.

Golubkova, K., Pinchuk, D. and Lopatka, J. (2015) "Bypassing Ukraine will be Costly for Russia's Gazprom – Analysts." *Reuters*, July 16, at: www.reuters.com/article/2015/07/16/ukraine-crisis-gazprom-idUSL5N0ZV1KM20150716 (accessed July 23, 2015).

Gorshkov, D. (2015) "Eyeing Russia, US Leads Fresh Military Drills in Ukraine." *Agence France-Presse*, July 20, at: www.france24.com/en/20150720-eyeing-russia-us-leads-fresh-military-drills-ukraine (accessed July 21, 2015).

Graziadei, V. (2014) "The Cunning Demons of Russian Propaganda, or What the BBC Forgot to Warn Us About." *Wordpress.com*, June 25, at: http://veragraziadei.wordpress.com/2014/06/25/the-cunning-demons-of-russian-propaganda-or-what-the-bbc-forgot-to-warn-us-about (accessed July 16, 2014).

Grey, B. (2014) "Ukrainian Military Kills Russian Journalists in Assault on Rebel-Held Town." *World Socialist Web Site*, June 18, at: www.wsws.org/en/articles/2014/06/18/ukra-j18.html (accessed August 11, 2014).

Grozev, C. (2015) "Quantum of Obfuscation." *Bellingcat.com*, October 16, at: www.bellingcat.com/news/uk-and-europe/2015/10/16/quantum-of-obfuscation/comment-page-2 (accessed November 30, 2015).

Grytsenko, O. (2015) "Mariupol Makes Second Attempt to Elect Mayor after Failed Vote." *Kyiv Post*, November 29, at: www.kyivpost.com/article/content/ukraine-politics/

mariupol-makes-second-attempt-to-elect-mayor-after-failed-vote-403114.html (accessed December 17, 2015).

Gunnar, U. (2014) "MH17: Malaysia's Barring from Investigation Reeks of Cover-Up." *New Eastern Outlook*, November 28, at: http://journal-neo.org/2014/11/28/mh17-malaysia-s-barring-from-investigation-reeks-of-cover-up/ (accessed June 29, 2016).

Harding, L. (2016) "How to Hide a Billion Dollars." *Guardian*, April 5, at: www.theguardian.com/news/2016/apr/03/what-you-need-to-know-about-the-panama-papers (accessed June 29, 2016).

Hart, P. (2014) " 'Muddying a Murky Picture': NYT's Ukraine Kremlinology." *FAIR.org*, June 3, at: www.fair.org/blog/2014/06/03/muddying-a-murky-picture-nyts-ukraine-kremlinology (accessed July 17, 2014).

Hartog, E. (2015) "Russian Media, Officials Continue Assault on Dutch MH17 Report." *Moscow Times*, October 14 at: www.themoscowtimes.com/news/article/russian-media-officials-continue-assault-on-dutch-mh17-report/539129.html (accessed December 21, 2015).

Head, M. (2015) "Russia Vetoes MH17 Tribunal Push by US Allies." *World Socialist Web Site*, July 30, at: www.wsws.org/en/articles/2015/07/30/mh17-j30.html (accessed August 31, 2015).

Helmer, J. (2014) "Ukraine Takes Another $1.39 Billion from International Monetary Fund–$3 Billion in IMF Cash Already Sent Offshore–Insiders Suspected in Heist." *Dances with Bears*, September 10, at: http://johnhelmer.net (accessed September 3, 2014).

Helmer, J. (2015a) "The Lie to End all Truths, and New Evidence." *Naked Capitalism*, September 19, at: www.nakedcapitalism.com/2015/09/john-helmer-mh17-the-lie-to-end-all-truths-and-new-evidence.html (accessed December 17, 2015).

Helmer, J. (2015b) "Australian Police, Dutch Investigators Break with Dutch Safety Board at First Coroner's Inquest into MH17 Crash." *Naked Capitalism*, December 16, at: www.nakedcapitalism.com/2015/12/john-helmer-australian-police-dutch-prosecutors-break-with-dutch-safety-board-at-first-coroners-court-inquest-on-mh17-crash.html (accessed December 17, 2015).

Helmer, J. (2015c) "MH17 Coroner Contradicts Himself, Ignores Australian Forensic Evidence, Rules in Favour of Dutch Safety Board." *OpEdNews*, December 16, at: www.opednews.com/Quicklink/MH17-Coroner-Contradicts-H-by-Natylie-Baldwin-Australia_Dutch-Safety-Board_Mh17-Aircraft_Mh17-Airline-Crash-151217-367.html (accessed December 18, 2015).

Herman, E. and Chomsky, N. (2002) *Manufacturing Consent: The Political Economy of the Mass Media*. New York: Pantheon.

Herszenhorn, D. (2015a) "Russia Heightens Dispute with Ukraine's Gazprom over Natural Gas." *New York Times*, February 24, at: www.nytimes.com/2015/02/25/world/europe/russia-warns-ukraine-over-natural-gas-supply.html?_r=0 (accessed March 2, 2015).

Herszenhorn, D. (2015b) "Dutch Say Separatists in Ukraine Block Inquiry into Malaysia Jet Crash." *New York Times*, June 28, at: www.nytimes.com/2015/06/29/world/europe/dutch-say-inquiry-into-malaysia-jet-crash-is-being-blocked-in-ukraine.html?_r=0 (accessed June 28, 2015).

Heyden, U. (2015a) "West Ukraine is Now a Permanent Playground for NATO Forces." *Russia Insider*, November 26, at: http://russia-insider.com/en/no-end-nato-maneuvers-west-ukraine/ri11502 (accessed January 7, 2016).

Heyden, U. (2015b) "Ukraine, When the Right Sector Runs the Courtroom." *Counterpunch*, December 18, at: www.counterpunch.org/2015/12/18/ukraine-when-the-right-sector-runs-the-courtroom (accessedJanuary 8, 2016).

Hewitt, K. (2015) "What is Going on in Russia? The Views and Values of Ordinary Russians." *OpEdNews.com*, December 17, at www.opednews.com/Quicklink/What-is-going-on-in-Russia-by-Natylie-Baldwin-Media-Distortion_Media-Lies_Media-Manipulation_Propaganda-151218-726.html (accessed June 20, 2016).

Higgins, A. (2015) "In Ukraine, It's Putin's Game." *New York Times*, February 11, at: www.nytimes.com/2015/02/12/world/in-ukraine-its-putins-game.html?_r=0, (accessed February 12, 2015).

Higgins, A. and Gordon, M.R. (2015) "Russians Open Fire in Ukraine, NATO Reports." *New York Times*, August 22, at: www.nytimes.com/2014/08/23/world/europe/russian-convoy-ukraine.html (accessed May 6, 2015).

Higgins, A. and Kramer, A. (2015) "Ukraine Leader Was Defeated Even Before He Was Ousted." *New York Times*, January 3, at: www.nytimes.com/2015/01/04/world/europe/ukraine-leader-was-defeated-even-before-he-was-ousted.html (accessed January 21, 2015).

Hirst, M. (2015) "If You Can't Make It Fake It." *Sputnik News*, February 27, at: http://sputniknews.com/europe/20150219/1018476918.html (accessed February 27, 2015).

Hitchens, P. (2014) "Who is Using Spies and Lies to Grab Power in Kiev? We Are." *Daily Mail*, April 19, at: www.dailymail.co.uk/debate/article-2608679/PETER-HITCHENS-Who-using-spies-lies-grab-power-Kiev-We-are.html (accessed June 20, 2016).

Hitchens, P. (2015) "A (Not So) Brief History of Crimea." *Mail on Sunday*, February 21, at: http://hitchensblog.mailonsunday.co.uk/2015/02/a-not-so-brief-history-of-crimea.html (accessed February 27, 2015).

Hudson, M. (2014a) "The New Cold War's Ukraine Gambit." In S. Lendman (ed.), *Flashpoint in Ukraine*. Atlanta, GA: Clarity Press, pp. 26–45.

Hudson, M. (2014b) "The IMF's New Cold War Loan to Ukraine." *Counterpunch*, September 9, at: www.counterpunch.org/2014/09/09/the-imfs-new-cold-war-loan-to-ukraine (accessed September 13, 2014).

Hudson, M. (2015a) "The Russian Loan and the IMF's One-Two Punch." *Counterpunch*, February 26, at: www.counterpunch.org/2015/02/16/ukraine-denouement (accessed March 14, 2015).

Hudson, M. (2015b) "The IMF Changes its Rules to Isolate China and Russia." *Counterpunch*, December 20, at: www.counterpunch.org/2015/12/18/the-imf-changes-its-rules-to-isolate-china-and-russia (accessed January 6, 2016).

Hugues, S. (2015) "French Officials Debunk NATO Warnings of Russian Invasion of Ukraine." *World Socialist Web Site*, April 14, at: www.wsws.org/en/articles/2015/04/14/fren-a14.html (accessed April 22, 2015).

HRW (2014) "A Day in Luhansk: War's Crimes, Horrors, and Uncertainties." *HRW*, July 5, at: www.hrw.org/news/2014/07/05/day-Luhansk-war-s-crimes-horrors-and-uncertainties (accessed August 11, 2014).

Hyland, J. and Mitchell, P. (2010) "UN Court Upholds Kosovo's Declaration of Independence." *World Socialist Web Site*, July 30, at: www.wsws.org/en/articles/2010/07/koso-j30.html (accessed January 9, 2015).

Interfax (2014) "Self-Proclaimed Donetsk People's Republic Leader denies Destruction of Russian Armor by Ukrainian Govt Forces." *Interfax*, August 16, at: www.interfax.com/newsinf.asp?id=529040 (accessed June 29, 2016).

JIT (Joint Investigation Team) (2016) "MH17: Six Aspects of the Criminal Investigations Explained." *JIT*, at: om.nl/mh17-ezine-juni2016/e-zine-en.html (accessed June 29, 2016).

Johnson, R. (2014) "Media on Ukraine: What Happened to Journalism?" *OpEdNews.com*, May 5, at: www.opednews.com/articles/Media-on-Ukraine-What-Hap-by-Reginald-Johnson-Media_Media-Bias_Media-Distortion_Media-Failure-140505-509.html (accessed June 18, 2014).

Jowett, G. and O'Donnell, V. (2015) *Propaganda and Persuasion*. Thousand Oaks, CA: Sage.

Judah, N. (2014) "Ukraine: Hate in Progress." *New York Review of Books*, April 28, at: www.nybooks.com/blogs/nyrblog/2014/apr/28/ukraine-hate-progress/ (accessed June 26, 2014).

Karatnycky, A. (2015) "Putin's Warlords Slip Out of Control." *New York Times*, June 9, at: www.nytimes.com/2015/06/10/opinion/putins-warlords-slip-out-of-control.html?_r=0 (accessed June 11, 2015).

Karlin, A. (2015) "Trump is Factually Right on Putin and Journalists." *UNZreview.com*, December 27, at: www.unz.com/akarlin/trump-right-on-putin (accessed January 7, 2016).

Karmanau, Y. and Leonard, P. (2014) "There's increasing evidence that pro-Russian rebels shot down the plane." *Business Insider*, July 25, available online at www.businessinsider.com/mh17-story-2014-7 (accessed July 19, 2016).

Katchanovski, I. (2015a) "Crimea: People and Territory Before and After Annexation." In A. Pikulicka-Wilczewska and Richard Sakwa (eds.), *Ukraine and Russia: People, Politics, Propaganda and Perspectives*. Bristol, UK: E-International Relations Publishing, pp. 80–89.

Katchanovski, I. (2015b) "The 'Snipers' Massacre' on the Maidan in Ukraine." *Annual Meeting of American Political Science Association*, San Francisco, September 3–6.

Katchanovski, I. (2015c) *The "Snipers' Massacre" on the Maidan in Ukraine*. University of Ottawa, September 5at http://papers.ssrn.com/sol3/papers.cfm?abstract_id=2658245 (accessed June 29, 2016).

Kaufman, L. (Dir) (2016) *Maidan: Tonight and Tomorrow*. Reel Peak Films.

Kelly, A. (2015) "One Year Later, Crimeans Prefer Russia." *Truthdig.com*, February 6, at: www.truthdig.com/eartotheground/item/one_year_later_crimeans_prefer_russia_20150206?utm_source=feedburner&utm_medium=feed&utm_campaign=Feed%253A+Truthdig+Truthdig%253A+Drilling+Beneath+the+Headlines (accessed February 10, 2015).

Kimberley, M. (2015) "War is Peace in Ukraine." *Black Agenda Report*, February 12, at: www.blackagendareport.com/node/14661 (accessed March 6, 2015).

Klare, M. (2016) "The Oil Pricequake." *Le Monde Diplomatique* (English edition), January 13, at: http://mondediplo.com/openpage/the-oil-pricequake (accessed January 13, 2016).

Klikushin, M. (2015) "Ukraine Spirals into the Abyss." *Observer*, July 23, at: http://observer.com/2015/07/ukraine-spirals-into-the-abyss-pensioner-suicides-and-open-talk-of-default (accessed July 4, 2016).

Kosaretsky, K. (2015) "German Sociologists on Crimea's Choice." *Oriental Review.org.*, February 10, at: http://orientalreview.org/2015/02/10/german-sociologists-on-crimeas-choice (accessed June 29, 2016).

Koshiw, J.V. (2015) *MH17: The Story of the Shooting Down of the Malaysian Airliner*. Charleston, SC: Createspace Independent Publishing Platform.

Kramer, A. (2014a) "Ukraine Vote on Separation Held in Chaos." *New York Times*, May 11, at: www.nytimes.com/2014/05/12/world/europe/ukraine-referendum.html (accessed June 20, 2016).

Kramer, A. (2014b) "A Ukraine City Under Siege, 'Just Terrified of the Bombing.'" *New York Times*, August 14, at: www.nytimes.com/2014/08/15/world/europe/a-ukraine-city-under-siege-just-terrified-of-the-bombing.html?_r=0 (accessed August 19, 2014).

Kramer, A. (2015a) "Mindful of Greece, Ukraine is in a Rush to Line Up Debt Relief." *New York Times*, July 4, at: www.nytimes.com/2015/07/04/business/mindful-of-greece-ukraine-is-in-a-rush-to-line-up-debt-relief.html?_r=0 (accessed July 22, 2015).

Kramer (2015b) "Islamic Battalions, Stocked With Chechens, Aid Ukraine in War With Rebels." *New York Times*, July 7, at: www.nytimes.com/2015/07/08/world/europe/islamic-battalions-stocked-with-chechens-aid-ukraine-in-war-with-rebels.html?_r=0 (accessed July 4, 2016).

Kredo, A. (2015) "Updated: Provenance of Photos Showing Atrocities Questioned." *Washington Free Beacon*, February 12, at: http://freebeacon.com/national-security/exclusive-photos-show-russian-military-in-ukraine-arming-separatists (accessed February 17, 2015).

Krieger, M. (2014) "Ukraine's Finance Minister: 'Made in the USA': Kiev Government Giving Away Citizenships so 'Foreigners Can Run the Country.'" *Global Research*, December 6, at: www.globalresearch.ca/ukrainian-elites-made-in-the-usa-kiev-government-giving-away-citizenships-so-foreigners-can-run-the-country/5417956 (accessed January 21, 2015).

Kristof, N. (2014) "In Dad's Hometown, European Souls." *New York Times*, April 19, at: www.nytimes.com/2014/04/20/opinion/sunday/in-dads-hometown-european-souls.html?_r=0 (accessed June 20, 2016).

Kuzio, T. (2015) "The Origins of Peace, Non-Violence, and Conflict in Ukraine." In A. Pikulicka-Wilczewska and Richard Sakwa (eds.), *Ukraine and Russia: People, Politics, Propaganda and Perspectives*. Bristol, UK: E-International Relations Publishing, pp. 109–122.

Lantier, A. (2014a) "The NYT Finds Russian Spies in Eastern Ukraine." *World Socialist Web Site*, April 22, at: www.wsws.org/en/articles/2014/04/22/time-a22.html?view=article_mobile (accessed August 11, 2014).

Lantier, A. (2014b) "NYT Tries to Whitewash Publication of Faked Ukraine Photos." *World Socialist Web Site*, April 24, at: www.wsws.org/en/articles/2014/04/24/nytu-a24.html (accessed June 26, 2014).

Lantier, A. (2014c) "US-Backed Regime Retakes Slavyansk, Threatens Bloodbath in Eastern Ukraine." *World Socialist Web Site*, July 7, at: www.wsws.org/en/articles/2014/07/07/ukra-j07.html (accessed August 11, 2014).

Lantier, A. (2014d) "Netherlands Covers Up Secret Pact Controlling Investigation of MH17 Crash." *World Socialist Web Site*, November 22, at: www.wsws.org/en/articles/2014/11/22/mh17-n22.html (accessed December 15, 2014).

Lantier, A. (2015a) "German, French Leaders Fly to Russia Amidst Warnings of 'Total War' over Ukraine." *World Socialist Web Site*, February 6, at: www.wsws.org/en/articles/2015/02/06/ukra-f06.html (accessed February 10, 2015).

Lantier, A. (2015b) "Europe on the Brink of War." *World Socialist Web Site*, February 7, at: www.wsws.org/en/articles/2015/02/07/pers-f07.html (accessed February 7, 2015).

Lantier, A. (2015c) "IMF Loans $17 Billion to Ukraine to Repay Banks, Loot Economy." *World Socialist Web Site*, March 13, at: www.wsws.org/en/articles/2015/03/13/imfu-m13.html (accessed March 13, 2015).

Laruelle, M. (2015) "The Ukrainian Crisis and its Impact on Transforming Russian Nationalism Landscape." In A. Pikulicka-Wilczewska and Richard Sakwa (eds.), *Ukraine and Russia: People, Politics, Propaganda and Perspectives*. Bristol, UK: E-International Relations Publishing, pp. 123–128.

Laclau, E. and Mouffe, C. (2014) *Hegemony and Socialist Strategy: Towards a Radical Democratic Politics*. London: Verso.

Lebedev, V. (2014) "Sergey Lavrov: The Truth must be Revealed." *Russia Direct*, August 26, at: www.russia-direct.org/russian-media/sergey-lavrov-truth-must-be-revealed (accessed June 29, 2016).

Lendman, S. (2014) "Introduction." In S. Lendman (ed.), *Flashpoint in Ukraine*. Atlanta, GA: Clarity Press, pp. 9–14.

Lendman, S. (2015) "Nemtsov Murder: Anti-Putin False Flag." *Institute for Political Economy*, February 28, at: www.paulcraigroberts.org/2015/02/28/nemtsov-murder-anti-putin-false-flag-stephen-lendman (accessed February 28, 2015).

Leonard, P. (2014) "Ukraine: Air Force Jet Downed by Russian Missile." *Yahoo.com*, July 17, at: http://news.yahoo.com/ukraine-air-force-jet-downed-russian-missile-102817809.html (accessed January 23, 2016).

Leurs, R. (2014) "Chief MH17 Investigator on German Claims: 'We Will Need Evidence.'" *Spiegel Online International*, October 2, at: www.spiegel.de/international/world/mh17-chief-investigator-says-no-actionable-evidence-yet-in-probe-a-999485.html (accessed December 1, 2014).

Lisitsyn, P. (2014) "Reports of Russia's Military Build-Up on Ukraine Border Groundless – Moscow." *RIA Novosti*, August 6, at: http://rt.com/news/178444-russia-military-ukraine-nato (accessed August 11, 2014).

Litvinova, D. (2015) "Russian Foreign Minister Accuses Netherlands of Concealing Facts About MH17." *Moscow Times*, August 24, at: www-cf.themoscowtimes.com/news/article/russian-foreign-minister-accuses-netherlands-of-concealing-facts-about-mh17/528545.html (accessed August 31, 2015).

Lloyd, R. and Postol, T. (2014a) "Whose Sarin?" *London Review of Books*, Vol. 36, No. 10, May 22, at: www.lrb.co.uk/v36/n10/letters (accessed June 29, 2016).

Lloyd, R. and Postol, T. (2014b) "Possible Implications of Faulty US Technical Intelligence in the Damascus Nerve Agent Attack of August 21, 2013." MIT Science, Technology, and Global Security Working Group.

Losh, J. (2015a) "Ukraine Crisis: Soldiers Accuse Commanders of Lying as Both Warring Sides make Repeated Claims of Victories and Broken Ceasefires." *The Independent*, August 19, at: www.independent.co.uk/news/world/europe/ukraine-crisis-soldiers-accuse-commanders-of-lying-as-both-warring-sides-make-repeated-claims-of-victories-and-broken-ceasefires-10462754.html (accessed August 30, 2015).

Losh, J. (2015b) "Did the Islamic State Go Shopping for Weapons in Ukraine." *Vice News*, December 1, at: https://news.vice.com/article/did-the-islamic-state-go-shopping-for-weapons-in-ukraine?utm_source=vicenewsemail (accessed December 2, 2015).

Lough, J. (2015) "Gazprom is Facing a Perfect Storm." *Moscow Times*, November 27, at: www.themoscowtimes.com/opinion/article/gazprom-is-facing-a-perfect-storm-op-ed/550847.html (accessed December 8, 2015).

Lubin, G. (2014) "How Russians Became Crimea's Largest Ethnic Group, in One Haunting Chart." *Business Insider*, March 16, at: www.businessinsider.com/crimea-demographics-chart-2014-3 (accessed January 15, 2016).

Lyachynska, O. (2016) "No Revolution Here: News Media In Ukraine Remain In Same Hands." *Kyiv Post*, January 14, at: www.kyivpost.com/article/content/ukraine-politics/no-revolution-here-news-media-in-ukraine-remain-in-same-hands-405942.html (accessed January 19, 2016).

Lynch, J. and McGoldrick, A. (2005) *Peace Journalism*. Stroud, Glos: Hawthorne Press.

Maavak, M. (2014) "MH17, Gaza and the 'Genius' of Western Propaganda." *Pravda*,

August 21, at: http://english.pravda.ru/world/americas/21-08-2014/128338-western_propaganda-0 (accessed September 2, 2014).

McAdams, D. (2014) "US Turns Blind Eye to Lugansk Massacre." *Ron Paul Institute*, June 4, at: https://ronpaulinstitute.org/archives/featured-articles/2014/june/04/us-turns-blind-eye-to-Lugansk-massacre.aspx (accessed August 11, 2014).

McAdams, D. (2015) "'Minsk II' – What about Foreign Troops in Ukraine?" *Ron Paul Institute*, February 12, at: www.ronpaulinstitute.org/archives/peace-and-prosperity/2015/february/12/minsk-ii-what-about-foreign-troops-in-ukraine (accessed March 6, 2015).

McCombs, M. (2014) *Setting the Agenda: Mass Media and Public Opinion*. London: Polity.

MacFarquhar, N. (2015) "Early Memo Urged Moscow to Annex Crimea, Report Says." *New York Times*, February 26, A4.

McGovern, R. (2014a) "How NATO jabs Russia on Ukraine." *Consortiumnews.com*, May 15, at: http://consortiumnews.com/2014/05/15/how-nato-jabs-russia-on-ukraine (accessed June 29, 2014).

McGovern, R. (2014b) "Russia's Humanitarian 'Invasion.'" *Consortiumnews.com*, August 23, at: http://consortiumnews.com/2014/08/23/russias-humanitarian-invasion (accessed August 28, 2014).

McGovern, R. (2015) "Propaganda, Intelligence and MH-17." *Consortiumnews.com*, August 17, at: https://consortiumnews.com/2015/08/17/propaganda-intelligence-and-mh-17 (accessed August 31, 2015).

Malic, N. (2015) "MSM Spin Theory that Putin Hijacked MH370 and Landed it in Kazakhstan." *Antiwar.com*, February 28, at: http://original.antiwar.com/malic/2015/02/27/msm-spins-theory-that-putin-hijacked-mh370-and-landed-it-in-kazakhstan (accessed March 1, 2015).

Margolis, E. (2015) "The Russians Are Coming." *Lewrockwell.com*, September 8, at: www.lewrockwell.com/2015/09/eric-margolis/the-russians-are-coming-the-russians-are-coming (accessed December 20, 2015).

Mamon, M. (2015a) "In Midst of War, Ukraine Becomes Gateway for Jihad." *The Intercept*, February 26, at: https://firstlook.org/theintercept/2015/02/26/midst-war-ukraine-becomes-gateway-europe-jihad (accessed July 22, 2015).

Mamon, M. (2015b) "ISA Munayev's War: The Final Days of a Chechen Commander Fighting in Ukraine." *The Intercept*, February 27, at: https://theintercept.com/2015/02/27/isa-munayevs-war/ (accessed June 20, 2016).

Marshall, J. (2016) "Nazi Roots of Ukraine's Conflict." *Consortiumnews.com*, January 28, at: https://consortiumnews.com/2016/01/28/nazi-roots-of-ukraines-conflict (accessed February 3, 2016).

Martenson, C. (2015) "The Road To War With Russia. We're Not Only on It; We've Already Arrived." *Information Clearing House*, January 16, at: www.informationclearinghouse.info/article40721.htm (accessed March 3, 2015).

Martin, P. (2015) "Danger Mounts of US/NATO War with Russia." *World Socialist Web Site*, August 13, at: www.wsws.org/en/articles/2015/08/13/pers-a13.html (accessed August 31, 2015).

Maybury, G. (2015) "Propaganda v Impropaganda – MH17, Brand 60 Minutes and Shell-Game Journalism." *OpEdNews.com*, May 25, at: www.opednews.com/articles/Propaganda-v-Impropaganda-by-Greg-Maybury-60-Minutes_Blowback_International_Investigations-150525-589.html (accessed June 4, 2015).

Mazneva, E. (2015) "Ukraine President Says Obama Approved Sending New Weapons."

*Bloomberg.com*, October 4, at: www.bloomberg.com/news/articles/2015-10-04/ukraine-president-says-obama-approved-sending-new-weapons (accessed December 8, 2015).

Meijer, R. (2015) "When Did We All Become Murderers? The West's Appalling Evolution from Supporter of Democracy to Supporter of Death Squads." *Russia Insider*, April 26, at: http://russia-insider.com/en/when-did-we-all-become-murderers/6056 (accessed April 27, 2015).

Melanovski, J. (2015) "Ukrainian Government Planning to Move Forward with Mass Privatizations." *World Socialist Web Site*, November 13, at: www.wsws.org/en/articles/2015/11/13/ukra-n13.html (accessed December 24, 2015).

Mercouris, A. (2015a) "IMF Dithers as Ukraine Teeters on Defaulting on Russia Loan." *TheRussioPhile.org*, December 18, at: www.therussophile.org/imf-dithers-as-ukraine-teeters-on-defaulting-on-russia-loan.html (accessed January 6, 2016).

Mercouris, A. (2015b) "Ukraine's Local Elections Expose Poroshenko's Weakening Grip." *Russia Insider*, December 27, at: russia-insider.com/en/politics/ukraines-local-elections-expose-poroshenkos-weakening-grip/ri10762 (accessed December 23, 2015).

Mercouris, A. (2016) "Assessing a Murder Case Against Putin." *Consortiumnews.com*, January 27, at: https://consortiumnews.com/2016/01/27/assessing-a-murder-case-against-putin (accessed February 3, 2016).

Meyssan, T. (2015) "The Russian Army is Beginning to Engage in Syria: Cuba Inside the World." August 26, at: https://cubainsidetheworld.wordpress.com/2015/08/26/the-russian-army-is-beginning-to-engage-in-syria (accessed August 31, 2015).

Mikhailov, K. (2015) "More SBU Information on Russian Aggression to be Released." *Kyiv Post*, July 3, at: www.kyivpost.com/content/kyiv-post-plus/more-sbu-information-on-russian-aggression-to-be-released-392661.html (accessed July 15, 2015).

Mokrushyna, H. (2015a) "Families Describe Fleeing War-Torn Ukraine." *Truthout*, May 24, at: www.truth-out.org/news/item/30934-families-describe-fleeing-war-torn-ukraine-for-russian-camps (accessed June 9, 2015).

Mokrushyna, H. (2015b) "Decentralization Reform in Ukraine." *Counterpunch*, August 28, at: www.counterpunch.org/2015/08/28/decentralization-reform-in-ukraine (accessed December 23, 2015).

Mokrushyna, H. (2015c) "On Ukraine's 'Incorrect' Past." *Counterpunch*, October 7, at: www.counterpunch.org/2015/10/09/on-ukraines-incorrect-past (accessed December 23, 2015).

Molden, P. (2014) *Understanding Priming Effects in Social Psychology*. New York: The Guilford Press.

Monroe, D. (2014) "Derek Monroe: The Fog of War brings US Satellite Images Allegedly Showing Russia Firing Rockets into Ukraine into Question." *The Real News Network*, July 28, at: http://therealnews.com/t2/index.php?option=com_content&task=view&id=31&Itemid=74&jumival=12169 (accessed August 6, 2014).

Moreira, P. (Dir) (2016) *Ukraine: Masks of the Revolution*. Premières Lignes Télévision.

*Moscow Times* (2015a) "Russia's Arctic Drilling Rig Pumps Over 4 Million Barrels of Crude." *Moscow Times*, July 30, at: www.themoscowtimes.com/business/article/russias-arctic-drilling-rig-pumps-over-4-million-barrels-of-crude/526516.html (accessed August 31, 2015).

*Moscow Times* (2015b) "Putin Admits Military Presence in Ukraine During Marathon Presser." *Moscow Times*, December 17, at: www.themoscowtimes.com/news/article/putin-admits-military-presence-in-ukraine-during-marathon-presser/553173.html (accessed December 18, 2015).

Mousseau, F. (2015) "The Corporate Takeover of Ukrainian Agriculture." *Inter Press*

*Service*, January 27, at: www.ipsnews.net/2015/01/opinion-the-corporate-takeover-of-ukrainian-agriculture (accessed February 7, 2015).

Nazemroaya, M.D. (2014a) "The Geo-Politics of Euromaidan." In Stephen Lendman (ed.), *Flashpoint in Ukraine*. Atlanta, GA: Clarity Press, pp. 90–108.

Nazemroaya, M.D. (2014b) "Mapping a New World Order: The World After World War III." *Information Clearing House*, September 11, at: www.informationclearinghouse.info/article39638.htm (accessed September 13, 2014).

Nazemroaya, M.D. (2015) "ISIS-Linked Jihadists-for-Hire Now Active in Ukraine." *Russia Insider*, at: http://russia-insider.com/en/politics/isil-using-ukraine-forward-base-caucasia-and-entry-europe/6412 (accessed July 22, 2015).

Nelson, J. (2014) "GM Food, Ukraine, and the Return of Hill + Knowlton." *GM Watch*, August 22–24, at: http://gmwatch.org/index.php/news/archive/2014/15609-gm-food-ukraine-and-the-return-of-hill-knowlton (accessed September 2, 2014).

*New York Times* Editorial Board (2014a) "What Mr. Putin Can't Control." *New York Times*, May 12, at: www.nytimes.com/2014/05/13/opinion/what-mr-putin-cant-control.html (accessed June 20, 2016).

*New York Times* Editorial Board (2014b) "A Critical Election in Ukraine." *New York Times*, May 21, at: www.nytimes.com/2014/05/21/opinion/a-critical-election-in-ukraine.html?_r=0 (accessed June 20, 2016).

*New York Times* Editorial Board (2015) "Russia's Empty Gesture on the Downed Malaysian Jet." *New York Times*, July 23, at: www.nytimes.com/2015/07/23/opinion/russias-empty-gesture-on-the-downed-malaysian-jet.html?_r=0 (accessed July 24, 2015).

Niemöller, J. (2014) *MH17: de Doofpotdeal*. Amsterdam: Uitgeverij Van Praag.

Noack, R. (2014) "A Worrying Factor in Ukraine's Chaos: Fifteen Nuclear Reactors." *Washington Post*, September 3, at: http://news.nationalpost.com/2014/09/03/a-worrying-factor-in-ukraines-chaos-fifteen-nuclear-reactors (accessed September 6, 2014).

Noble, S. (2012) *The Power Principle, Parts 1 (Empire) and 2 (Propaganda)*. Metanoia Films, at: http://metanoia-films.org/the-power-principle (accessed January 1, 2016).

North, D. (2015) "Murder in Moscow: Why Was Boris Nemtsov Assassinated?" *World Socialist Web Site*, March 3 at: www.wsws.org/en/articles/2015/03/03/pers-m03.html (accessed March 3, 2015).

Oakford, S. (2015) "Russia's Economy is a Mess – and Its Problems Aren't Going Away." *Vice News*, September 1, at: https://news.vice.com/article/russias-economy-is-a-mess-and-its-problems-arent-going-away (accessed December 21, 2015).

Oakland Institute (2014) "The World Bank and the IMF Open Up Ukraine to Western Interests." *Oakland Institute*, July 28, at: www.oaklandinstitute.org/press-release-world-bank-and-imf-open-ukraine-western-interests (accessed September 13, 2014).

Ogirenko, V. (2015) "Russia Expresses 'Deep Concern' Over Ukraine's Decentralization Plan." *Moscow Times*, July 3, at: www.themoscowtimes.com/news/article/russia-expresses-deep-concern-over-ukraines-decentralization-plan/524955.html (accessed July 21, 2015).

Orlov, D. (2015) "On the 19th Day of Christmas." *ClubOrlov*, December 22, at: www.cluborlov.blogspot.com/2015/12/on-19th-day-of-christmas.html (accessed January 7, 2016).

Parry, R. (2014a) "Putin or Kerry: Who's Delusional?" *Consortiumnews.com*, March 5, at: http://consortiumnews.com/2014/03/05/putin-or-kerry-whos-delusional (accessed August 11, 2014).

Parry, R. (2014b) "Mainstream US Media Is Lost in Ukraine." *Consortiumnews.com*, March 16, at: http://consortiumnews.com/2014/03/16/mainstream-us-media-is-lost-in-ukraine/ (accessed June 18, 2014).

Parry, R. (2014c) "Evidence for Russian Involvement in East Ukraine Based on Shoddy Journalism." *The Real News Network*, April 23, at: http://therealnews.com/t2/index.php?option=com_content&task=view&id=31&Itemid=74&jumival=11773 (accessed June 20, 2016).

Parry, R. (2014d) "Twisting Putin's Words on Ukraine." *Consortiumnews.com*, May 2, at: http://consortiumnews.com/2014/05/02/twisting-putins-words-on-ukraine (accessed August 11, 2014).

Parry, R. (2014e) "Will Ukraine Be NYT's Waterloo?" *Consortiumnews.com*, May 3, at: http://consortiumnews.com/2014/05/03/will-ukraine-be-nyts-waterloo (accessed August 11, 2014).

Parry, R. (2014f) "Another NYT 'Sort of' Retraction on Ukraine." *Consortiumnews.com*, May 4, at: http://consortiumnews.com/2014/05/04/another-nyt-sort-of-retraction-on-ukraine/ (accessed June 26, 2014).

Parry, R. (2014g) "Fascism Is Rearing Its Ugly Head in Ukraine." *Alternet*. May 6, at: www.alternet.org/world/ukraine-catastrophe-offers-many-frightening-reminders-nazi-era (accessed June 20, 2016).

Parry, R. (2014h) "Burning Ukraine's Protesters Alive." *Consortiumnews.com*, May 10, at: http://consortiumnews.com/2014/05/10/burning-ukraines-protesters-alive/ (accessed June 29, 2014).

Parry, R. (2014i) "Ethnic Russians Are People, Too." *Consortiumnews.com*. May 13, at: http://consortiumnews.com/2014/05/13/ethnic-russians-are-people-too/ (accessed June 20, 2016).

Parry, R. (2014j) "The State Department's Ukraine Fiasco." *Consortiumnews.com*, May 24, at: http://consortiumnews.com/2014/05/24/the-state-departments-ukraine-fiasco (accessed March 4, 2016).

Parry, R. (2014k) "NYT's One-Sided Ukraine Narrative." *Consortiumnews.com*, May 26, at: http://consortiumnews.com/2014/05/26/nyts-one-sided-ukraine-narrative/ (accessed June 26, 2014).

Parry, R. (2014l) "NYT Revamps Its False Ukraine Narrative." *Consortiumnews.com*, June 26, at: http://consortiumnews.com/2014/06/26/nyt-revamps-its-false-ukraine-narrative (accessed August 7, 2014).

Parry, R. (2014m) "NYT Dishes More Ukraine Propaganda." *Consortiumnews.com*, July 6, at: http://consortiumnews.com/2014/07/06/nyt-dishes-more-ukraine-propaganda (accessed August 7, 2014).

Parry, R. (2014n) "Airline Horror Spurs New Rush to Judgement." *Consortiumnews.com*, July 19, at: http://consortiumnews.com/2014/07/19/airline-horror-spurs-new-rush-to-judgment/ (accessed August 5, 2014).

Parry, R. (2014o) "The Mystery of a Ukrainian Army 'Defector.'" *Consortiumnews.com*, July 22, at: http://consortiumnews.com/2014/07/22/the-mystery-of-a-ukrainian-army-defector (accessed August 10, 2014).

Parry, R. (2014p) "Flight 17 Shoot-Down Scenario Shifts." *Consortiumnews.com*, August 3, at: http://consortiumnews.com/2014/08/03/flight-17-shoot-down-scenario-shifts (accessed August 10, 2014).

Parry, R. (2014q) "NYT Discovers Ukraine's Neo-Nazis at War." *Consortiumnews.com*, August 10, at: http://consortiumnews.com/2014/08/10/nyt-discovers-ukraines-neo-nazis-at-war (accessed August 11, 2014).

Parry, R. (2014r) "The Powerful 'Group Think' on Ukraine." *Consortiumnews.com*, August 18, at: https://consortiumnews.com/2014/08/18/the-powerful-group-think-on-ukraine/ (accessed June 20, 2016).

Parry, R. (2014s) "Malaysia Airlines Whodunnit Still a Mystery." *Consortiumnews.com*, September 9, at: http://consortiumnews.com/2014/09/09/malaysia-airlines-whodunnit-still-a-mystery (accessed September 10, 2014).

Parry, R. (2015a) "The Danger of an MH-17 'Cold Case.'" *Consortiumnews.com*, January 19, at: consortiumnews.com/2015/01/19/the-danger-of-an-mh-17-cold-case (accessed March 4, 2016).

Parry, R. (2015b) "Crimeans Keep Saying No to Ukraine." *Consortiumnews.com*, March 22, at: https://consortiumnews.com/2015/03/22/crimeans-keep-saying-no-to-ukraine (accessed March 24, 2015).

Parry, R. (2015c) "Gifting Russia 'Free-Market' Extremism." *Consortiumnews.com*, May 5, at: https://consortiumnews.com/2015/05/05/gifting-russia-free-market-extremism (accessed June 7, 2015).

Parry, R. (2015d) "A Reckless 'Stand-Upper' on MH17." *Consortiumnews.com*, May 28, at" https://consortiumnews.com/2015/05/28/a-reckless-stand-upper-on-mh-17 (accessed June 7, 2015).

Parry, R. (2015e) "Neocon Fugitive Given Ukraine Province." *Consortiumnews.com*, June 2, at: https://consortiumnews.com/2015/06/02/neocon-fugitive-given-ukraine-province (accessed July 8, 2015).

Parry, R. (2015f) "Cold War II to McCarthyism II." *Consortiumnews.com*, June 8, at: https://consortiumnews.com/2015/06/08/cold-war-ii-to-mccarthyism-ii (accessed June 11, 2015).

Parry, R. (2015g) "NYT's Orwellian View of Ukraine." *Consortiumnews.com*, June 22, at: https://consortiumnews.com/2015/06/22/nyts-orwellian-view-of-ukraine (accessed July 16, 2015).

Parry, R. (2015h) "MH-17 Case Slips Into Propaganda Fog." *Consortiumnews.com*, July 9, at" https://consortiumnews.com/2015/07/09/mh-17-case-slips-into-propaganda-fog (accessed July 16, 2015).

Parry, R. (2015i) "The Power of False Narrative." *Consortiumnews.com*, September 28, at: https://consortiumnews.com/2015/09/28/the-power-of-false-narrative (accessed December 20, 2015).

Parry, R. (2015j) "MH-17 Case: 'Old' Journalism vs. 'New.'" *Consortiumnews.com*, October 20, at: https://consortiumnews.com/2015/10/20/mh-17-case-old-journalism-vs-new (accessed December 17, 2015).

Parry, R. (2015k) "Carpetbagging 'Crony Capitalism' in Ukraine." *Consortiumnews.com*, November 13 at: https://consortiumnews.com/2015/11/13/carpetbagging-crony-capitalism-in-ukraine (accessed December 2, 2015).

Parry, R. (2015l) "Tangled Threads of US False Narratives." *Consortiumnews.com*, November 19, at: https://consortiumnews.com/2015/11/19/tangled-threads-of-us-false-narratives (accessed December 20, 2015).

Parry, R. (2016a) "MH-17's Unnecessary Mystery." *Consortiumnews.com*, January 15, at: https://consortiumnews.com/2016/01/15/mh-17s-unnecessary-mystery/?print=print (accessed June 29, 2016).

Parry, R. (2016b) "Kerry Pressed for MH-17 Evidence." *Consortiumnews.com*, January 21, at: https://consortiumnews.com/2016/01/21/kerry-pressed-for-mh-17-evidence (accessed January 23, 2016).

Parry, R. (2016c) "From Brady to MH-17: Power Defines Reality." *Consortiumnews.com*,

April 26, at: https://consortiumnews.com/2016/04/26/from-brady-to-mh-17-power-defines-reality (accessed June 29, 2016).

Parry, R. (2016d) "More Game-Playing on MH17?" *Consortiumnews.com*, May 24, at: https://consortiumnews.com/2016/05/24/more-game-playing-on-mh-17 (accessed June 29, 2016).

Parsons, R. (2014a) "US Media Follows White House Lead on Ukraine." *Counterpunch*, May 8, at: www.counterpunch.org/2014/05/08/us-media-follows-white-house-lead-on-ukraine (accessed June 27, 2014).

Parsons, R. (2014b) "More NATO False Claims: Ukraine Humanitarian Aid Flounders." *Global Research*, August 20, at: www.globalresearch.ca/more-nato-false-claims-ukraine-humanitarian-aid-flounders/5396701 (accessed September 2, 2014).

Paul, C. (2011) *Strategic Communication: Origins, Concepts and Current Debates*. Santa Barbara, CA: ABC-CLIO.

Paul, R. (2014a) "What the Media Won't Report About Malaysian Airlines Flight MH17." *Ron Paul Institute*, July 20, at: http://ronpaulinstitute.org/archives/featured-articles/2014/july/20/what-the-media-won%E2%80%99t-report-about-malaysian-airlines-flight-mh17.aspx (accessed July 21, 2014).

Paul, R. (2014b) "Reckless Congress 'Declares War' on Russia." *Ron Paul Institute*, December 4, at: www.ronpaulinstitute.org/archives/featured-articles/2014/december/04/reckless-congress-declares-war-on-russia/ (accessed June 20, 2016).

Perkins, A. (2005) *Confessions of an Economic Hit Man*. New York: Plume.

Peters, A. (2014) "Leaders of Pro-Russian Separatist Forces in Eastern Ukraine Resign." *World Socialist Web Site*, August 18, at: www.wsws.org/en/articles/2014/08/18/sepa-a18.html (accessed August 29, 2014).

Peters, A. (2015a) "Far-right Marches in Kiev as Report Documents Crimes by Ukraine's Paramilitary Forces." *World Socialist Web Site*, July 6, at: www.wsws.org/en/articles/2015/07/06/ukra-j06.html (accessed July 15, 2015).

Peters, A. (2015b) "PBS Documentary Putin's Way: Half-Truths and Lies in the Service of US Warmongering against Russia." *World Socialist Web Site*, September 9, at: www.wsws.org/en/articles/2015/09/09/puti-s09.html (accessed December 20, 2015).

Peterson, D. (2014) "U.S. Media Non-Coverage of the Urmas Paet – Catherine Ashton Sniper Revelation (Update)." *Z Blogs*, March 10, at: http://zcomm.org/zblogs/u-s-media-non-coverage-of-the-urmas-paet-catherine-ashton-sniper-revelation-update/ (accessed June 20, 2016).

Petras, J. (2014) "Obama's Ukrainian Power Grab." In Stephen Lendman (ed.), *Flash-point in Ukraine*. Atlanta, GA: Clarity Press, pp. 46–49.

Petras, J. (2015) "Putin and the Press: The Demonology School of Journalism." *Aletho News*, at: https://alethonews.wordpress.com/2015/10/12/putin-and-the-press-the-demonology-school-of-journalism (accessed December 20, 2015).

Petro, N. (2014) "Ukraine's Ongoing Struggle with its Russian Identity." *World Politics Review*, May 6, pp. 4–10.

Pilger, J. (2014) "The Return of George Orwell and Big Brother's War on Palestine, Ukraine and the Truth." *Johnpilger.com*, July 11, at: http://johnpilger.com/articles/the-return-of-george-orwell-and-big-brothers-war-on-palestine-ukraine-and-truth (accessed August 8, 2014).

Pogrebinskiy, M. (2015) "Russians in Ukraine: Before and after Euromaidan." In A. Pikulicka-Wilczewska and Richard Sakwa (eds.), *Ukraine and Russia: People, Politics, Propaganda and Perspectives*. Bristol, UK: E-International Relations Publishing, pp. 90–99.

Polk, W. (2015) "Pushing the Edge on Nuclear War." *Consortiumnews.org*, August 28, at: https://consortiumnews.com/2015/08/28/pushing-the-edge-on-nuclear-war (accessed August 31, 2015).
Porter, G. (2014) *Manufactured Crisis: The Untold Story of the Iran Nuclear Scare*. Charlottesville, VA: Just World Books.
Prentice, A. (2015) "Ukraine Health System in Danger of Collapse as Reforms Stall." *Reuters*, December 23, at: http://news.yahoo.com/ukraine-health-system-danger-collapse-reforms-stall-080421210-business.html (accessed January 7, 2016).
Putin, V. (2014) "Press Statements and Replies to Journalists' Questions." Kremlin, Moscow, May 7, at: http://en.kremlin.ru/events/president/transcripts/20973 (accessed June 20, 2016).
Queally, J. (2014) "NATO Claims Satellite Images Show Russian Units in Ukraine." *CommonDreams*, August 28, at: www.commondreams.org/news/2014/08/28/nato-claims-satellite-images-show-russian-units-ukraine (accessed June 29, 2016).
Rachkevych, M. (2015) "Ukraine's Massive Shadow Economy Robs Treasury and Future of 44 Million Citizens." *Kyiv Post*, December 27, at: www.kyivpost.com/article/content/business/ukraines-massive-shadow-economy-robs-treasury-and-future-of-44-million-citizens-404982.html (accessed January 2, 2016).
Raimondo, J. (2015a) "The Murderers of Kiev: Ukraine's Gangster Regime Shows its True Colors." *Antiwar.com*, April 16, at: http://original.antiwar.com/justin/2015/04/16/the-murderers-of-kiev (accessed April 22, 2015).
Raimondo, J. (2015b) "Monsters of the Ukraine: Made in the USA." *Antiwar.com*, July 31, at: https://antiwar.liberty.me/monsters-of-ukraine-made-in-the-usa (accessed August 31, 2015).
Raimondo, J. (2015c) "Ukraine's 'Democratic' Dictatorship." *Antiwar.com*, September 20, at: http://original.antiwar.com/justin/2015/09/20/ukraines-democratic-dictatorship (accessed December 24, 2015).
Rapoza, K. (2015) "One Year after Russia Annexed Crimea, Locals Prefer Moscow to Kiev." *Forbes*, March 20, at: www.forbes.com/sites/kenrapoza/2015/03/20/one-year-after-russia-annexed-crimea-locals-prefer-moscow-to-kiev (accessed March 24, 2015).
Rasmus, J. (2014) "Who Benefits from the Ukraine Economic Crisis?" In S. Lendman (ed.), *Flashpoint in Ukraine*. Atlanta, GA: Clarity Press, pp. 119–130.
Rasmus, J. (2015) "The New Colonialism: Greece and Ukraine." *Counterpunch*, September 3, at: www.counterpunch.org/2015/09/03/the-new-colonialism-greece-and-ukraine (accessed December 8, 2015).
Ray, J. (2015) "Gallup Poll: Ukrainians Disillusioned with Leadership." *Gallup*, December 23, at: www.gallup.com/poll/187931/ukrainians-disillusioned-leadership.asp (accessed January 7, 2016).
Ray, J. and Esipova, N. (2014) "Russian Approval of Putin Soars to Highest Level in Years." *Gallup*, July 18, at: www.gallup.com/poll/173597/russian-approval-putin-soars-highest-level-years.aspx (accessed January 19, 2016).
Reuters (2015a) "Russia's Gazprom Says Can Exempt Rebel-Held Areas from Ukraine Gas Contract." *Reuters.com*, February 26, at: www.reuters.com/article/2015/02/26/us-russia-crisis-ukraine-gas-idUSKBN0LU11M20150226 (accessed March 2, 2015).
Reuters (2015b) "Villagers Tell Their Story on MH17 Downing over Eastern Ukraine." *Moscow Times*, March 14, at: www.themoscowtimes.com/news/article/villagers-tell-their-story-on-mh17-downing-over-eastern-ukraine/517406.html (accessed March 17, 2015).
Reuters (2015c) "Captured Russian Soldier 'Admits' on Video to Spying in Eastern

Ukraine." *Moscow Times*, May 18, www.themoscowtimes.com/news/article/captured-russian-soldier-admits-on-video-to-spying-in-eastern-ukraine/520980.html (accessed June 7, 2015).

Reuters (2015d) "Battle Rages for Marinka." *UNIAN*, June 3, at: www.unian.info/war/1084840-battle-rages-for-marinka-25-ukrainian-soldiers-reported-wounded.html

Reuters (2015e) "Ukraine's Right Sektor Challenges Poroshenko after Police Standoff." *Moscow Times*, July 13, at: www.themoscowtimes.com/news/article/ukraines-right-sector-challenges-poroshenko-after-fatal-standoff-with-police/525473.html (accessed July 14, 2015).

Reuters (2015f) "Sanctions Cut German Exports to Russia by 34 Percent." *Kyiv Post*, July 23, at: www.kyivpost.com/content/russia-and-former-soviet-union/reuters-sanctions-cut-german-exports-to-russia-by-34-percent-394159.html

Reuters (2015g) "Dutch Exports to Russia Down by 40 Percent Post-MH17, Data Shows." *Moscow Times*, July 24, at: static.themoscowtimes.com/business/article/dutch-exports-to-russia (accessed August 30, 2015).

Reuters (2015h) "Ukraine's Gross Foreign Debt Grows to Almost $127 bln in Q2." *UNIAN Information Agency*, September 18, at: www.unian.info/economics/1128247-ukraines-gross-foreign-debt-grows-to-almost-to-127-bln-in-q2.html (accessed December 14, 2015).

Reuters (2015i) "Decline in Ukraine's Industrial Production in August Slows to 5.8%." *UNIAN.info*, September 21, at: www.unian.info/economics/1129216-decline-in-ukraines-industrial-production-in-august-slows-to-58.html (accessed December 6, 2015).

Reuters (2015j) "Ukraine Grain Exports at 14 Mln T So Far in 2015/16 – Ministry." *Reuters*, November 3, at: http://af.reuters.com/article/commoditiesNews/idAFL8N12Y4F120151103 (accessed November 24, 2015).

Reuters (2015k) "Bulgaria, Greece Sign Deal to Start Building Gas Pipeline." *Reuters.com*, December 10, at: www.reuters.com/article/bulgaria-greece-pipeline-idUSL8N13X28R20151210 (accessed January 6, 2016).

Reuters (2015l) "Ukraine Lawmaker Manhandles Prime Minister, Sparking Parliament Brawl." *NBC News*, December 11, at: www.nbcnews.com/storyline/ukraine-crisis/ukraine-lawmaker-manhandles-prime-minister-sparking-parliament-brawl-n478836 (accessed January 6, 2016).

RFE/RL (2015a) "Council of Europe Blasts Ukraine's Investigations into Odesa Violence." *RFE/RL*, November 4, at: www.rferl.org/content/ukraine-odesa-fire-council-europe-report/27345601.html (accessed December 3, 2015).

RFE/RL (2015b) "Saakashvili Slams Georgian Decision To Revoke His Citizenship." *RFE/RL*, December 24, at: www.rferl.org/content/georgia-saakashvili-citzenship-stripped/27406996.html (accessed December 23, 2015).

Roberts, P. (2014a) "Washington's Recklessness Endangers the World." In S. Lendman (ed.), *Flashpoint in Ukraine*. Atlanta, GA: Clarity Press, pp. 15–18.

Roberts, P. (2014b) "Washington has Set the World on a Path to War." In S. Lendman (ed.), *Flashpoint in Ukraine*. Atlanta, GA: Clarity Press, pp. 19–25.

Roberts, P. (2014c) "What Happened to the Malaysian Airliner?" *Paul Craig Roberts*, July 17, at: www.paulcraigroberts.org/2014/07/19/happened-malaysian-airliner-paul-craig-roberts (accessed August 11, 2014).

Roberts, P. (2014d) "Ukraine: In The West Respect for Truth No Longer Exists." *Foreign Policy Journal*, August 18, at: www.foreignpolicyjournal.com/2014/08/18/ukraine-in-the-west-respect-for-truth-no-longer-exists (accessed August 19, 2014).

Roberts, P. (2014e) "How Can You Tell Whether Russia Has Invaded Ukraine?" *OpEdNews*, September 1, at: www.opednews.com/articles/How-Can-You-Tell-Whether-R-by-Paul-Craig-Roberts-Disinformation_NATO_Propaganda_Russia-140901-822.html (accessed September 4, 2014).

Roberts, P. (2015) "Nemtsov's Assassination: A Propaganda Attack On Putin?" *Institute for Political Economy*, February 27, at: www.paulcraigroberts.org/2015/02/27/nemtsovs-assassination-propaganda-attack-putin (accessed June 29, 2016).

Robinson, J. (2016) "The Outbreak of Nuclear War Within a Year Between the West and Putin is 'Entirely Plausible', Says Former NATO Chief (While Promoting his Novel about a 2017 War with Russia)." *Daily Mail*, May 18, at: www.dailymail.co.uk/news/article-3596977/The-outbreak-nuclear-war-year-West-Putin-entirely-plausible-says-former-NATO-chief-promoting-novel-2017-war-Russia.html#ixzz4D1u5gl5r (accessed June 29, 2016).

Rombach, A. (2014) "Popular Discontent Grows with German Media Lies in Ukraine Crisis." *Global Research*, April 4, at: www.globalresearch.ca/popular-discontent-grows-with-german-media-lies-in-ukraine-crisis-2/5376623 (accessed June 29, 2014).

Roosevelt, K. (1979) *Counter Coup: The Struggle for Control of Iran*. New York: McGraw Hill.

Rosenberg, P. (2014) "Seven Decades of Nazi Collaboration: America's Dirty Little Ukraine Secret." *Foreign Policy In Focus*, March 28, at: http://fpif.org/seven-decades-nazi-collaboration-americas-dirty-little-ukraine-secret (accessed August 11, 2014).

Rozoff, R. (2014) "The Geostrategic Significance of Ukraine in NATO's Drive to the East." In Stephen Lendman (ed.), *Flashpoint in Ukraine*. Atlanta, GA: Charity Press, pp. 114–118.

RT (2014a) "Facts You Need to Know about Crimea and Why it is in Turmoil." *RT.com*, February 26, at: www.rt.com/news/crimea-facts-protests-politics-945 (accessed January 15, 2016).

RT (2014b) "OSCE Calls on Kiev to Free Russian Journalists, 'Stop Intimidating Media.'" *RT.com*, May 19, at: http://rt.com/news/159960-osce-kiev-russian-journalists/ (accessed June 18, 2014).

RT (2014c) "Malaysia Excluded from MH17 Probe – For 'Not Pointing Fingers at Russia.'" *RT.com*, November 28, at: http://rt.com/news/209759-mh17-investigation-malaysia-excluded (accessed December 15, 2014).

RT (2015a) "50,000 Casualties in Ukraine? German Intel Calls Kiev's 6k Toll 'Not Credible.'" *RT.com*, February 8, at: http://rt.com/news/230363-ukraine-real-losses-german-intelligence/ (accessed February 11, 2015).

RT (2015b) "Putin in Film on Crimea: US Masterminds behind Ukraine Coup, Helped Train Radicals." *RT.com*, March 15, at: www.rt.com/news/240921-us-masterminds-ukraine-putin (accessed July 24, 2015).

RT (2015c) "Ukraine Interior Ministry Was 'Uncooperative and Obstructive' in Maidan Crimes Probe – EU Report." *RT.com*, March 31, at: http://rt.com/news/245557-ukraine-maidan-probe-eu (accessed April 2, 2015).

RT (2015d) "'Unprecedented & Dangerous Step': Russia Slams NATO Troop Build-Up." *RT.com*, April 2, at: http://rt.com/news/246321-russia-nato-troops-border (accessed April 27, 2015).

RT (2015e) "'Why Do They keep Shooting?' Violence Spikes in E. Ukraine, OSCE blames Kiev Forces." *RT.com*, April 15, at: http://rt.com/news/249937-osce-ukraine-report-ceasefire (accessed April 27, 2015).

RT (2015f) "Israeli-Made Air-to-Air Missile May Have Downed MH17 – Report." *RT.com*, July 16, at: www.rt.com/news/310039-mh17-israeli-missile-version (accessed July 23, 2015).

RT (2015g) "Around 200 Masked Men Storm Kharkov City Hall in Eastern Ukraine." *RT.com* September 23, at: www.rt.com/news/316266-masked-men-storm-kharkov/ (accessed November 30, 2015).

RT (2015h) "Putin: All Countries Should Respect Ukraine's Sovereignty." *RT.com*, September 27, at: www.rt.com/news/316687-putin-interview-sovereignty-ukraine (accessed November 30, 2015).

RT (2015i) "Russia Powers Up Crimea, Cutting Dependence on Ukraine." *RT.com*, December 3, at: www.rt.com/business/324550-crimea-energy-bridge-russia (accessed December 21, 2015).

RT (2016a) " 'Biased, Low Quality, Full of Omissions': Russia Launches Fresh Attack on Dutch MH17 Report." *RT.com*, January 19, at: www.rt.com/news/329494-mh17-investigation-biased-omissions (accessed January 21, 2016).

RT (2016b) "Dutch Investigators Say No Sat Images of MH17 Crash Exist, Enquiry Could Last Years." *RT.com*, February 19, at: http://rt.com/news/333025-dutch-investigator-mh17-letter (accessed February 19, 2016).

Rudenko, O. (2015) "Election Wins Qualified International Praise." *Kyiv Post*, October 26, at: www.kyivpost.com/content/ukraine/election-wins-qualified-international-praise-ruling-party-loses-support-400702.html (accessed December 20, 2015).

Rudin, M. (Dir) (2016) *The Conspiracy Files: Who shot down MH17?* BBC.

*Russia Insider* (2015) "Why Kiev Silenced Ukraine's Biggest Opposition Newspaper." *Russia Insider*, October 3, at: http://russia-insider.com/en/politics/why-kiev-silenced-ukraines-biggest-opposition-newspaper/ri10696 (accessed December 8, 2015).

Saker, The (2015) "Ukraine: If Poroshenko Attacks his Days are Numbered." *Information Clearing House*, July 20, at: www.informationclearinghouse.info/article42435.htm (accessed July 23, 2015).

Sakwa, R. (2014) "The Ukraine Crisis and the Death of Europe." *TELOSscope*, September 15, at: www.telospress.com/the-ukraine-crisis-and-the-death-of-europe/ (accessed June 20, 2016).

Sakwa, R. (2015) *Frontline Ukraine: Crisis in the Borderlands*. London and New York: I.B. Tauris.

Sandhu, S. (2015) "MH17 Crash: Missile Fragments Reportedly 'Found in Bodies' of Victims: The Fragments are Believed to be from a Buk Missile." *Independent*, October 8, at: www.independent.co.uk/news/world/europe/mh17-crash-missile-fragments-reportedly-found-in-bodies-of-victims-a6685596.html (accessed December 17, 2015).

Savage, C. (2015) *Power Wars*. New York: Little, Brown and Co.

Schneider, H. (2015) "IMF Aims for 'Immediate' Stabilization with Latest Ukraine Bailout Deal." *Reuters*, March 11, at: www.reuters.com/article/2015/03/11/us-ukraine-crisis-imf-idUSKBN0M71D620150311 (accessed March 14, 2015).

Schofield, M. (2014) "New Analysis of Rocket used in Syria Chemical Attack Undercuts US Claims." *Star-Telegram*, January 16, at: www.star-telegram.com/2014/01/15/5488779/new-analysis-of-rocket-usedin.html?storylink=addthis#.Ute6sbgLS0s.twitter&rh=1 (accessed January 19, 2015).

Schrad, M. (2015) "Will There be Justice for MH17?" *Moscow Times*, June 28, at: www.themoscowtimes.com/article/524533.html (accessed July 20, 2015).

Schwarz, P. (2014a) "German President Gauck threatens Russia with War." *World Socialist Web Site*, September 3, at: www.wsws.org/en/articles/2014/09/03/gauc-s03.html (accessed September 4, 2014).

Schwarz, P. (2014b) "G-20 Summit Intensifies Threat of War against Russia." *World Socialist Web Site*, November 18, at: www.wsws.org/en/articles/2014/11/18/pers-n18.html (accessed January 22, 2015).

Schwarz, P. (2015) "Leading German Newspaper Says Russia Should be Threatened with Nuclear War." *World Socialist Web Site*, February 12, at: www.wsws.org/en/articles/2015/02/12/nuke-f12.html (accessed March 17, 2015).

Sengupta, S. (2015) "Vote by U.N. General Assembly isolates Russia." *New York Times*, March 27, at: www.nytimes.com/2014/03/28/world/europe/General-Assembly-Vote-on-Crimea.html (accessed January 2, 2015).

Shamir, I. (2014) "The Ukrainian Pendulum." *Counterpunch.org*, March 8, at: www.counterpunch.org/2014/03/07/the-ukrainian-pendulum (accessed July 22, 2014).

Shetty, S. (2014) "Ukraine: Mounting Evidence of War Crimes and Russian Involvement." *Amnesty International*, September 7, at: www.amnesty.org/en/news/ukraine-mounting-evidence-war-crimes-and-russian-involvement-2014-09-05 (accessed September 10, 2014).

Shoemaker, P. and Vos, T. (2009) *Gatekeeping Theory*. New York: Routledge.

Shukla, A. (2015) "Yugoslav Break-Up was Planned in Advance." *OpEdNews*, October 24, at: www.opednews.com/articles/Yugoslavia-break-up-was-p-by-Ashish-Shukla-Bombing_Disinformation_Energy_Germany-151024-446.html (accessed January 13, 2016).

Silino, Y. and Shylenko, O. (2015) "In Key Concession, Ukraine Rebels Delay Polls until 2016." *Agence France-Presse*, October 6, at: http://news.yahoo.com/ukraine-rebels-delay-local-elections-until-2016-002037271.html (accessed December 6, 2015).

Smith, P.L. (2014) "The New York Times Manufactures Ignorance: More Half-Truths about Ukraine." *Salon.com*. March 20, at: www.salon.com/2014/03/20/the_new_york_times_manufactures_ignorance_more_half_truths_about_ukraine/ (accessed June 20, 2016).

Smith, P.L. (2015a) "The New York Times 'Basically Rewrites Whatever the Kiev Authorities Say': Stephen F. Cohen on the U.S./Russia/Ukraine History the Media Won't Tell You." *Salon.com*, April 16, at: www.salon.com/2015/04/16/the_new_york_times_basically_rewrites_whatever_the_kiev_authorities_say_stephen_f_cohen_on_the_u_s_russiaukraine_history_the_media_wont_tell_you (accessed April 20, 2015).

Smith, P.L. (2015b) "We Are the Propagandists: The Real Story about How The New York Times and the White House has turned Truth in the Ukraine on its Head." *Salon.com*, June 3, at: www.salon.com/2015/06/03/we_are_the_propagandists_the_real_story_about_how_the_new_york_times_and_the_white_house_has_turned_truth_in_the_ukraine_on_its_head (accessed June 9, 2015).

Sokol, S. (2015) "Ukrainian Parliament Recognizes Militia that Collaborated with Nazis." *Jerusalem Post*, April 13, at: www.jpost.com/Diaspora/Ukrainian-parliament-recognizes-militia-that-collaborated-with-Nazis-396848 (accessed April 27, 2015).

*Der Spiegel* (2015) "Breedlove's Bellicosity: Berlin Alarmed by Aggressive NATO Stance on Ukraine." *Der Spiegel*, March, 7, at: www.spiegel.de/international/world/germany-concerned-about-aggressive-nato-stance-on-ukraine-a-1022193.html (accessed June 29, 2016).

*Sputnik News* (2015ba) "Guardian Corrects Anti-Russian Story based on 'Research' of Own Reporter." *Sputnik International*, February 19, at: http://sputniknews.com/military/20150219/1018504330.html (accessed June 29, 2016).

*Sputnik News* (2015b) "Right-Wing Terror in Kherson, Southern Ukraine Forces Residents to Take Up Arms." *Rightedition.com*, December 27, at: http://rightedition.com/2015/12/27/terror-in-southern-ukraine-forcing-residents-to-take-up-arms(accessed January 7, 2015).

*Sputnik News* (2016) "Study Proves Maidan Snipers Were Western-Backed Opposition's False Flag." *Sputnik International*, January 4, at: http://sputniknews.com/europe/20160103/1032633643/study-maidan-deaths-false-flag.html (accessed June 29, 2016).

Stanek, H. (2015) "The Forgotten Refugee Crisis in Ukraine." *OpEdNews*, October 2, at: www.opednews.com/articles/The-Forgotten-Refugee-Cris-by-Henry-Stanek-Donbas_Ukraine_Ukraine_Ukraine-Coup-151002-763.html (accessed December 8, 2015).

Steinberg, S. (2014) "Continuing Media Silence on the Fate of Flight MH17." *World Socialist Web Site*, August 29, at: www.wsws.org/en/articles/2014/08/29/mh17-a29.html (accessed September 6, 2014).

Stern, J. (2014) "German Television Report Indicates Ukrainian Opposition Responsible for Lethal Shootings." *World Socialist Web Site*, April 14, at: www.wsws.org/en/articles/2014/04/14/getv-a14.html (accessed June 20, 2016).

Stewart, W., Reuters, Associated Press and Gayle, D. (2014) "Kiev in U-turn over Claim that 'Russian Tanks, Artillery and 1,200 Fighters' Had Been Deployed in Eastern Ukraine as Evidence Fails to Materialise." *Daily Mail*, August 20, at: www.dailymail.co.uk/news/article-2729647/Tanks-artillery-1-200-fighters-march-east-Ukraine-Russia-days-Putin-set-meet-Ukrainian-president-peace-talks.html (accessed September 2, 2014).

Stockman, D. (2014) "On Dominoes, WMDs And Putin's 'Aggression': Imperial Washington is Intoxicated by Another Big Lie." *Contra Corner*, at: http://davidstockmanscontracorner.com/on-dominoes-wmds-and-putins-aggression-imperial-washington-is-intoxicated-in-another-big-lie (accessed August 11, 2014).

Stone, O. (Exec Prod) (2016) *Ukraine on Fire*. Another Way Productions.

Sukhov, O. (2015a) "Fears Rise over Adequacy of Mariupol's Defenses." *Kyiv Post*, May 28, at: www.kyivpost.com/content/ukraine/fears-rise-over-adequacy-of-mariupols-defenses-389769.html (accessed June 7, 2015).

Sukhov, O. (2015b) "Thousands Protest Voting Fraud in Kryvyi Rih, Call for Uprising." *Kyiv Post*, November 22, at: www.kyivpost.com/content/kyiv-post-plus/thousands-protest-over-voting-fraud-in-kryvyi-rih-call-for-uprising-402616.html (accessed December 14, 2015).

Tanchum, M. (2015) "ANAP Pipeline Project and the Semi-Encirclement of Iran: Progress and Paradoxes in Turkey's Energy Diplomacy." *Turkish Policy Quarterly*, December 28, at: www.turkishpolicy.com/article/773/the-future-of-the-balkans-is-in-europe-what-about-the-past-and-the-present (accessed January 6, 2016).

Tartakovsky, J. (2015) "Donetsk: A Defiant and Besieged City." *Truthout*, May 18, at: www.truth-out.org/speakout/item/30740-donetsk-a-defiant-and-besieged-city (accessed June 9, 2015).

TASS (2015) "16,000 Ukrainian Troops defect Donbas, Many with Weapons." *TASS*, October 5, at: http://tass.ru/en/world/826234 (accessed November 24, 2015).

Taub, A. (2015) "Pro-Kiev Militias are Fighting Putin, But has Ukraine Created a Monster it Can't Control?" *Vox*, February 23, at: www.vox.com/2015/2/20/8072643/ukraine-volunteer-battalion-danger (accessed Feburary 23, 2015).

Tikhonova, P. (2015) "Europe Turns Away from US Toward Russia." *Valuewalk.com*, November 4, at: www.valuewalk.com/2015/11/europe-turns-away-from-u-s-toward-russia (accessed December 6, 2015).

Tiezzio, S. (2014) "China's 'New Silk Road' Vision Revealed." *The Diplomat*, May 14, at: http://thediplomat.com/2014/05/chinas-new-silk-road-vision-revealed (accessed July 30, 2015).

Timtchenko, I. and Gordiienko, O. (2015) "After a Year of War against Ukraine, Russia is Still the Country's Leading Trade Partner." *Kyiv Post*, June 26, at: www.kyivpost.com/content/business/after-a-year-of-war-against-ukraine-russia-is-still-the-countrys-leading-trade-partner-392123.html (accessed July 14, 2015).

Transparency International (2016) "Corruption Perceptions Index 2015." *Transparency International*, at: www.transparency.org/cpi2015 (accessed June 29, 2016).

Troianovski, A. and Lee, C. (2014) "Mistrust Persists in Ukraine Meetings." *Wall Street Journal*, June 7, at: http://online.wsj.com/articles/russias-putin-and-ukraine-president-elect-poroshenko-meet-in-france-1402058620 (accessed June 26, 2014).

Uehling, G. (2015) "Everyday Life after Annexation: The Autonomous Republic of Crimea." In Agnieszka Pikulicka-Wilczewska and Richard Sakwa (eds.), *Ukraine and Russia: People, Politics, Propaganda and Perspectives*. Bristol, UK: E-International Relations Publishing, pp. 69–79.

Uhler, W.C. (2014a) "The Propaganda War over Ukraine: The New York Times vs. Russia's 'White Book': Part One." *Dissident Voice*. May 9, at: http://dissidentvoice.org/2014/05/the-propaganda-war-over-ukraine/ (accessed June 26, 2014).

Uhler, W.C. (2014b) "*New York Times* Disinformation Campaign over Ukraine: Part Two: The Propaganda War over Ukraine." *Dissident Voice*, June 12, at: http://dissidentvoice.org/2014/06/new-york-times-disinformation-campaign-over-ukraine/ (accessed June 26, 2014).

UNIAN (2015a) "Reuters: UN Reports More Signs of Russian Involvement in Donbass War." *UNIAN Information Agency*, June 1, at: www.unian.info/war/1084216-reuters-un-reports-more-signs-of-russian-involvement-in-Donbass-war.html (accessed July 21, 2015).

UNIAN (2015b) "Ambassador Chaly: A Dozen European Countries Arming Ukraine Now." *UNIAN Information Agency*, July 10, at: www.unian.info/politics/1099520-ambassador-chaly-a-dozen-european-countries-arming-ukraine-now.html (accessed July 13, 2015).

UNIAN (2015c) "Churkin: Russia Virtually Removed from MH17 Investigation." *UNIAN Information Agency*, July 16, at: www.unian.info/world/1101599-churkin-russia-virtually-removed-from-mh17-investigation.html (accessed June 29, 2016).

UNIAN (2015d) "Yarosh Wants Amnesty for Right Sector Fighters Involved in Mukacheve Conflict." *UNIAN Information Agency*, July 17, at: www.unian.info/politics/1102146-yarosh-wants-amnesty-for-right-sector-fighter-involved-in-mukacheve-conflict.html (accessed July 20, 2015).

UNIAN (2015e) "SCE Chairman Urges DPR Militants to Stop Attacks on Monitors." *UNIAN Information Agency*, August 19, at: www.unian.info/politics/1113277-osce-chairman-urges-dpr-militants-to-stop-attacks-on-monitors.html (accessed August 31, 2015).

UNIAN (2015f) "From Kremlin-armed Donbas Militants to Millionaire Criminal Bosses." *UNIAN Information Agency*, August 28, at: www.unian.info/economics/1116478-from-kremlin-armed-donbas-militants-to-millionaire-criminal-bosses.html (accessed August 31, 2015).

UNIAN (2015g) "Ukrainian Journalists Expose Fake 'Ukrainian' and 'US' Websites Managed from Russia." *UNIAN Information Agency*, September 4, at: www.unian.info/society/1118976-ukraine-today-ukrainian-journalists-expose-fake-ukrainian-and-us-websites-managed-from-russia.html (accessed December 6, 2015).

UNIAN (2015h) "Three Ukrainian Soldiers Detained in Crimea." *UNIAN Information Agency*, September 7, at: www.unian.info/politics/1119761-three-ukrainian-soldiers-detained-in-crimea.html (accessed December 8, 2015).

UNIAN (2015i) "Poll – Almost Half of Ukrainians Believe That No Party Represents Their Interests." *UNIAN Information Agency*, September 16, at: www.unian.info/society/1125142-poll-almost-half-of-ukrainians-believe-that-no-party-represents-their-interests.html (accessed December 17, 2015).

UNIAN (2015j) "Yatsenyuk Promises Increased Pensions as of October." *UNIAN Information Agency*, September 20, at: www.unian.info/economics/1129111-yatsenyuk-promises-increased-pensions-as-of-october.html (accessed December 20, 2015).

UNIAN (2015k) "Coal Production in Ukraine Halves Since Early 2015." *UNIAN Information Agency*, October 5, at: www.unian.info/economics/1143357-coal-production-in-ukraine-halves-since-early-2015.html (accessed December 2, 2015).

UNIAN (2015l) "Deficit of Ukraine's Balance of Payments Shrinks to $1 Mln." *UNIAN Information Agency*, October 10, at: www.unian.info/economics/1169557-deficit-of-ukraines-balance-of-payments-shrinks-to-1-mln.html (accessed June 20, 2016).

UNIAN (2015m) "Sales of Industrial Goods Grow by 24.4% Since Year's Start." *UNIAN Information Agency*, October 10, at: http://ukrainianwall.com/english-news/sales-of-industrial-goods-grow-by-24-4-since-years-start/ (accessed December 23, 2015).

UNIAN (2015n) "Average Wages in Ukraine Grow by $6 to equivalent of $190 in Sept." *UNIAN Information Agency*, October 25, at: www.unian.info/economics/1164841-average-wages-in-ukraine-grow-by-6-to-equivalent-of-190-in-sept.html (accessed November 30, 2015).

UNIAN (2015o) "ATO HQ Reports Serious Escalation in Donbas, Intelligence Detects Enemy Drills." *UNIAN Information Agency*, November 25, at: www.unian.info/war/1193297-ato-hq-reports-serious-escalation-in-donbas-intelligence-detects-enemy-preparations.html (accessed December 6, 2015).

UNIAN (2015p) "Ukraine-Russia Trade Turnover Halves in Nine Months of 2015." *UNIAN Information Agency*, November 16, at: www.unian.info/economics/1185532-ukraine-russia-trade-turnover-halves-in-nine-months-of-2015.html (accessed December 24, 2015).

UNIAN (2015q) "UN Reports on Human Rights Abuses in Russian-Occupied Crimea." *UNIAN Information Agency*, December 9, at: www.unian.info/society/1207738-un-reports-on-human-rights-abuses-in-russian-occupied-crimea.html (accessed January 7, 2016).

UNIAN (2015r) "Ukraine Sees Slower Decline in Industrial Production in November." *UNIAN Information Agency*, December 21, at: www.unian.info/economics/1218598-ukraine-sees-slower-decline-in-industrial-production-in-november.html (accessed January 4, 2016).

UNIAN (2016a) "Inflation in Ukraine Reaches 20-Year Peak." *UNIAN Information Agency*, January 6, at: www.unian.info/economics/1230367-inflation-in-ukraine-reaches-20-year-peak.html (accessed January 6, 2016).

UNIAN (2016b) "BILD: Russia Spends About EUR 79 Mln on Donbas Salaries, Pensions per Month." *UNIAN Information Agency*, January 17, at: www.unian.info/economics/1238890-bild-russia-spends-about-eur-79-mln-on-donbas-salaries-pensions-per-month.html (accessed January 17, 2016).

United Nations Human Rights (2015) "Close to 8,000 People Killed in Eastern Ukraine." Office of High Commissioner, September 8, at: www.ohchr.org/EN/NewsEvents/Pages/DisplayNews.aspx?NewsID=16394 (accessed June 29, 2016).

US-Ukraine Business Council (2015) *Ukraine-Macroeconomic Situation*. USUBC, November, at: www.usubc.org/site/recent-news/ukraine---macroeconomic-situation---november-2015 (accessed July 4, 2016).

Van Auken, B. (2014a) "New York Times' Vision of Ukrainian 'Democracy.'" *World Socialist Web Site*, May 23, at: www.wsws.org/en/articles/2014/05/23/nyti-m23.html (accessed June 26, 2014).

Van Auken, B. (2014b) "Gaza, Ukraine and the Fraud of 'Human Rights' Imperialism." *World Socialist Web Site*, August 1, at: www.wsws.org/en/articles/2014/07/31/pers-j31.html (accessed August 11, 2014).

Van Auken, B. (2015a) "The New York Times and Washington's 'Propaganda Bubble' on Ukraine." *World Socialist Web Site*, January 6, at: www.wsws.org/en/articles/2015/01/06/time-j06.html (accessed January 21, 2015).

Van Auken, B. (2015b) "Clash of Oligarchs Shakes Ukrainian Regime." *World Socialist Web Site*, March 26, at: www.wsws.org/en/articles/2015/03/26/ukra-m26.html (accessed March 26, 2015).

Van der Graaf (2015) "Dutch Newspaper: Corruption in Ukrainian Secret Service Taints MH17 Investigation." *OpEdNews.com*, December 21, at: www.opednews.com/Quicklink/Dutch-newspaper-Corruptio-by-Natylie-Baldwin-Dutch-Prosecutor_Dutch-Safety-Board_Mh17-Aircraft_Mh17-Airline-Crash-151221-259.html (accessed December 21, 2015).

Vlasova, A. (2015) "Claim of Ukraine Weapons Sale to ISIS Prompts Denials, Alarm." *The Sofia Globe*, November 20, at: http://sofiaglobe.com/2015/11/21/claim-of-ukrainian-weapons-sale-to-isis-prompts-denials-alarm (accessed December 2, 2015).

Vlemmix, P. (2015) "FLIGHT MH17: Why We Still Don't Know What Happened – Peter Vlemmix Keep it Real (E1)." *Youtube.com*, March 13, at: www.youtube.com/watch?v=Y5kTf6EAMBA (accessed June 4, 2015).

Voice of Russia (2014) "'White Book' of Human Rights' Violations in Ukraine is 'Invitation to Act' – Moscow." July 2, at: http://voiceofrussia.com/news/2014_07_02/White-Book-of-human-rights-violations-in-Ukraine-is-invitation-to-act-Moscow-0565 (accessed September 14, 2014).

Volchek, D. and Bigg, C. (2015) "Ukraine Bloggers use Social Media to Track Russian Soldiers Fighting in East." *Guardian*, June 2, at: www.theguardian.com/world/2015/jun/03/bloggers-social-media-russian-soldiers-fighting-in-ukraine (accessed July 22, 2015).

Vorobiov, I. (2015) "Ukraine is a Mess, But it's Still No Greece." *Foreign Policy*, July 6, at: http://foreignpolicy.com/2015/07/06/ukraine-is-a-mess-but-its-still-no-greece (accessed July 22, 2015).

Walker, S. (2014a) "Ukraine and Crimea: What is Putin thinking?" *Guardian*, March 23, at: www.theguardian.com/world/2014/mar/23/ukraine-crimea-what-putin-thinking-russia (accessed June 24, 2014).

Walker, S. (2014b) "Ukraine President Claims Russian Vehicles That Crossed Border Have Been Destroyed." *Guardian*, August 16, at: www.theguardian.com/world/2014/aug/15/ukraine-claims-destroyed-russian-military-vehicles-crossed-border (accessed June 29, 2016).

Walker, S. (2014c) "Azov Fighters Are Ukraine's Greatest Weapon and May Be its Greatest Threat." *Guardian*, September 10, at: www.theguardian.com/world/2014/sep/10/azov-far-right-fighters-ukraine-neo-nazis (accessed January 1, 2015).

Walker, S. (2015) "Moscow and Kiev in 'Positive Mood' over Talks to End East Ukraine Conflict." *Guardian*, October 2, at: www.theguardian.com/world/2015/oct/02/moscow-and-kiev-in-positive-mood-over-ending-east-ukraine-conflict (accessed December 20, 2015).

Walt, S. (2015) "Why Arming Kiev Is a Really, Really Bad Idea: Washington Pundits are Jumping on a Proposal to Send Weapons to Ukraine. Here's Why They All Need to Take a Deep Breath." *Foreign Policy*, February 9, at: http://foreignpolicy.com/2015/02/09/how-not-to-save-ukraine-arming-kiev-is-a-bad-idea/ (accessed February 11, 2015).

Wearden, G. (2014) "Oxfam: 85 Richest People as Wealthy as Poorest Half of the World." *Guardian*, January 20, at: www.theguardian.com/business/2014/jan/20/oxfam-85-richest-people-half-of-the-world (accessed January 18, 2016).

Weber, M. ([1904] 1949) *The Methodology of the Social Sciences*. Glencoe, IL: Free Press.

Weiss, C. (2015a) "European Energy Firms Agree to Expand Nord Stream Pipeline with Gazprom." *World Socialist Web Site*, June 30, at: www.wsws.org/en/articles/2015/06/30/russ-j30.html (accessed July 8, 2015).

Weiss, C. (2015b) "German–Russian Energy Deal Worth Billion." *World Socialist Web Site*, November 3, at: www.wsws.org/en/articles/2015/11/03/gazp-n03.html (accessed December 8, 2015).

Werth, C. and Williams C. (2014) "Dutch Report: Malaysia Jet Downed in Ukraine by 'High-Energy Objects.'" *Los Angeles Times*, September 9, at: www.latimes.com/world/europe/la-fg-malaysia-ukraine-crash-investigation-20140909-story.html (accessed September 10, 2014).

Wesolowsky, T. (2015) "Cybersleuth Points to Russian Tank Unit in Eastern Ukraine." *RFE/RL*, September 23, at: www.rferl.org/content/russia-ukraine-cybersleuth-askai707-tanks-bellingcat/27265175.html (accessed December 5, 2015).

Weymouth, L. (2014) "Ukrainian Internal Affairs Minister: The Separatists 'Will Be Punished Severely.'" *Washington Post*, April 21, at: www.washingtonpost.com/opinions/ukrainian-internal-affairs-minister-the-separatists-will-be-punished-severely/2014/04/21/d01a1576-c985-11e3-93eb-6c0037dde2ad_story.html (accessed June 18, 2014).

Whitney, M. (2014a) "False Flag in Odessa." *Counterpunch*, May 8, at: www.counterpunch.org/2014/05/08/false-flag-in-odessa (accessed March 4, 2016).

Whitney, M. (2014b) "News Blackout – US Media Ignores Putin's Peace Plan." *Global Research*, May 12, at: www.globalresearch.ca/news-blackout-the-u-s-media-ignores-putins-peace-plan/5381937 (accessed June 18, 2014).

Whitney, M. (2014c) "Et Tu, Mullah? Did Iran Just Knife Putin in the Back?" *OpEd News*, August 12, at: www.opednews.com/articles/Et-Tu-Mullah--Did-Iran-by-Mike-Whitney-Crisis_Iran_NATO_Pipeline-140818-553.html (access August 29, 2014).

Whitney, M. (2014d) "My Money's on Putin." *Counterpunch*, August 12, at: www.counterpunch.org/2014/08/12/my-moneys-on-putin (accessed August 28, 2014).

Whitney, M. (2014e) "Understanding Ukraine in 15 Minutes." *Counterpunch*, August 22–24, at: www.counterpunch.org/2014/08/22/understanding-ukraine-in-15-minutes (accessed August 28, 2014).

Whitney, M. (2014f) "The Downing of Malaysia Flight 17: Sinister Pretext for War with Russia." *Counterpunch*, September 4, at: www.counterpunch.org/2014/09/04/sinister-pretext-for-war-with-russia (accessed December 15, 2014).

Whitney, M. (2014g) "Did Putin Just Bring Peace to Ukraine?" *Counterpunch*, September 5, at: www.counterpunch.org/2014/09/05/did-putin-just-bring-peace-to-ukraine (accessed September 10, 2014).

Whitney, M. (2014h) "The Ukraine, As We Know It, is Gone Forever." *Counterpunch*, October 14, at: www.counterpunch.org/2014/10/14/the-ukraine-as-we-know-it-is-gone-forever (accessed January 1, 2015).

Whitney, M. (2014i) "Irreversible Decline." *Counterpunch*, December 29 at: www.counterpunch.org/2014/12/29/irreversible-decline (accessed March 3, 2015).

Whitney, M. (2015a) "Oil Price Blowback: Is Putin Creating a New World Order?" *Counterpunch*, January 6, at: www.counterpunch.org/2015/01/06/oil-price-blowback-is-putin-creating-a-new-world-order (accessed March 3, 2015).

Whitney, M. (2015b) "Another Idiotic Plan to Hurt Russia." *Counterpunch*, April 20, at: www.counterpunch.org/2015/04/20/another-idiotic-plan-to-hurt-russia (accessed April 22, 2015).

Williams, C. (2014) "Pro-Russia Separatists Target Key Assets in Eastern Ukraine." *Los Angeles Times*, October 7, at: www.latimes.com/world/europe/la-fg-ukraine-russia-targets-20141007-story.html (accessed January 2, 2015).

Williams, C. (2015a) "Ukraine Gets Major Debt Relief from Creditors – Except Russia." *Los Angeles Times*, August 27, at: www.latimes.com/world/europe/la-fg-ukraine-debt-relief-20150827-story.html (accessed August 31, 2015).

Williams, C. (2015b) "U.N. Reports Nearly 8,000 Deaths in Ukraine War, Laments Russian Role." *Los Angeles Times*, September 8, at: www.latimes.com/world/europe/la-fg-ukraine-russia-un-deaths-20150908-story.html (accessed December 29, 2015).

Williamson, N. (2014) "Kiev's Blocking of Russian Aid Convoy to East Ukraine Escalates Risk of War." *World Socialist Web Site*, August 16, at: www.wsws.org/en/articles/2014/08/16/ukwp-a16.html (accessed August 29, 2014).

Williamson, N. (2015a) "IMF Announces New $17 Billion Loan Agreement for Ukraine." *World Socialist Web Site*, February 13, at: www.wsws.org/en/articles/2015/02/13/ukra-f13.html (accessed March 4, 2015).

Williamson, N. (2015b) "Fraud, Cancellations, Mass Abstention Mark Ukrainian Local Elections." *World Socialist Web Site*, October 27, at: www.wsws.org/en/articles/2015/10/27/ukra-o27.html (accessed December 6, 2015).

Wilson, A. (2015) "Ukrainian Politics Since Independence." In A. Pikulicka-Wilczewska and Richard Sakwa (eds.), *Ukraine and Russia: People, Politics, Propaganda and Perspectives*. Bristol, UK: E-International Relations Publishing, pp. 102–108.

World Bank (2016) "Population, Total." *World Bank Data*, at: http://data.worldbank.org/indicator/SP.POP.TOTL (accessed January 11, 2016).

Woronczuk, A. (2014) "Evidence for Russian Involvement in East Ukraine Based on Shoddy Journalism." *Real News Network*, April 24, at: http://truth-out.org/news/item/23281-evidence-for-russian-involvement-in-east-ukraine-based-on-shoddy-journalism (accessed January 9, 2015).

Xie, S. and Boyd-Barrett, O. (2015) "External-National TV News Networks' Way to America: Is the USA Losing the Global 'Information War'?" *International Journal of Communication*, Vol. 9.

Xinhua (2015) "Rebels in Ukraine's Donetsk Plan Referendum on Joining Russia: Media." *Xinhuanet.com*, August 20, at: http://news.xinhuanet.com/english/2015-08/19/c_134535215.htm (accessed August 31, 2015).

Zuesse, E. (2014a) "The New York Review of Books as Propagandist on Ukraine's Civil War." OpEdNews.com, June 2, at: www.opednews.com/articles/The-New-York-Review-of-Boo-by-Eric-Zuesse-Bilderberger_Media-Distortion_Media-Hypocrisy_Media-Mainstream-MSM-140602-600.html (accessed June 26, 2014).

Zuesse, E. (2014b) "More Than 50,000 US-backed Troops Are Fighting in Ukraine's Civil War. Why?" OpEdNews, August 30, at: www.opednews.com/articles/More-than-50-000-U-S-back-by-Eric-Zuesse-Obama-Administration_President-Barack-Obama-POTUS_Russia_Ukraine-140830-956.html (accessed September 8, 2014).

Zuesse, E. (2014c) "Western News-Suppression about the Downing of MH-17 Malaysian Jet." *Washington's Blog*, November 5, at: www.washingtonsblog.com/2014/11/western-news-suppression-downing-mh-17-malaysian-jet.html (accessed December 15, 2014).

Zuesse, E. (2014d) "Head of Stratfor, 'Private CIA,' Says Overthrow of Yanukovych Was 'The Most Blatant Coup in History.'" *Global Research*, December 20, at: www.globalresearch.ca/head-of-stratfor-private-cia-says-overthrow-of-yanukovych-was-the-most-blatant-coup-in-history/5420978 (accessed February 10, 2015).

Zuesse, E. (2014e) "Another MH17 Cover-Up: Hiding a Key Autopsy." *The People's Voice*, December 26, at: www.thepeoplesvoice.org/TPV3/Voices.php/2014/12/26/another-mh17-cover-up-hiding-a-key-autop (accessed December 26, 2014).

Zuesse, E. (2015a) "Barack Obama vs. Vladimir Putin: How Bloomberg 'News' Lies." *Global Research*, January 5, at: www.globalresearch.ca/barack-obama-vs-vladimir-putin-how-bloomberg-news-lies/5422472 (accessed February 11, 2015).

Zuesse, E. (2015b) "The May 2nd Odessa Massacre: Why Obama's Coup-Regime Still Runs Ukraine." *Washington's Blog*, April 30, at: www.washingtonsblog.com/2015/04/the-may-2nd-odessa-massacre-why-obamas-coup-regime-still-runs-ukraine.html (accessed May 4, 2015).

Zverev, A. (2014) "Exclusive: Ukraine Rebel Commander acknowledges Fighters Had BUK Missile." *Reuters*, July 23, at: www.reuters.com/article/2014/07/23/us-ukraine-crisis-commander-exclusive-idUSKBN0FS1V920140723 (access August 11, 2015).

# Index

Printed by Amazon Italia Logistica S.r.l.
Torrazza Piemonte (TO), Italy

31306823R00125